LEXICOGRAPHICA Series Maior

LEXICOGRAPHICA

Series Maior

Supplementary Volumes to the International Annual for Lexicography
Suppléments à la Revue Internationale de Lexicographie
Supplementbände zum Internationalen Jahrbuch für Lexikographie

Edited by
Sture Allén, Pierre Corbin, Reinhard R. K. Hartmann,
Franz Josef Hausmann, Hans-Peder Kromann, Oskar Reichmann,
Ladislav Zgusta

31

Published in cooperation with the Dictionary Society of North America
(DSNA) and the European Association for Lexicography (EURALEX)

Turki Diab

Pedagogical Lexicography

A Case Study of Arab Nurses as Dictionary Users

Max Niemeyer Verlag
Tübingen 1990

CIP-Titelaufnahme der Deutschen Bibliothek

Pedagogical Lexicography : a case study of Arab nurses as dictionary users / ed. by Turki Diab.
– Tübingen : Niemeyer, 1990
(Lexicographica : Series maior ; 31)
NE: Diab, Turki [Hrsg.]; Lexicographica / Series maior

ISBN 3-484-30931-8 ISSN 0175-9264

Printed in Germany.
Druck: Weihert-Druck GmbH, Darmstadt

TABLE OF CONTENTS

LIST OF TABLES

CHAPTER ONE
INTRODUCTION

1.1 ESP: Theoretical Framework

1.1.1 Background

Educational authorities in many developing countries, Jordan included, have long realised the importance of accelerating the process of science and technology transfer; and as a consequence it is becoming widely accepted that English has an essential role to play in this process since a great deal of scientific and technological communication takes place through this language. In academic institutions, universities in particular, a good command of English has indeed become a prerequisite for success in academic as well as professional activities.

Within the domain of foreign language teaching, the rise of ESP (English for Specific Purposes) has been one of the most interesting and significant developments over the past few decades or so. It is not my intention in this introductory chapter to provide a comprehensive state-of-the-art survey of the ESP field. This is, on the one hand, beyond the scope of this work and, on the other hand, is retrievable from the published literature (see, e.g., Robinson 1980; Coffey 1984). Rather, I propose to highlight some of the major principles that are likely to have some bearing on the nature of this study.

ESP has emerged as a pragmatic response to the ineffective practices that prevailed in the teaching of English as a foreign language before the 1960's. Many practitioners in various teaching situations felt unhappy about the irrelevant and inappropriate materials, designs and methodologies at the time. According to McDonough (1984:4):

> "Beginning in the early 1960s, there were many reports from around the world of a growing dissatisfaction with the language teaching practice then current, where all learners were served up with literature regardless of their aims, needs, or interests."

A new and different approach to language teaching had therefore to be sought where the learners' 'aims, needs or interests' could be accommodated and met. ESP became a necessity when learners, mostly adults, no longer saw their aims of learning the foreign language in terms of pleasure seeking or in terms of appreciating a foreign culture. Rather, students wanted to learn the foreign language to achieve practical goals.

In institutions of tertiary education, the need for a new orientation in English language teaching was more strongly felt. The role of English had changed from the learner's point of view. That is to say, English was no longer needed as a study of the language for its own sake, but rather as a means to achieve 'utilitarian' purposes (Mackay and Mountford 1978:2). There were also changes in the economies of newly independent countries and in order to speed up their economic development, there were urgent needs to learn English, the language of communication in various sectors of the economy, industry and business (Strevens 1980:105).

Also, behind the rise of ESP were the following factors, put forward by Candlin (in Mackay and Mountford 1978:vi):

1. The introduction of the new concept of communicative need of the learner, the implication of which was a need to specify the purposes for which the foreign language was learned.
2. The increased interest in studying languages in social contexts in the 1960's.
3. The developments in linguistics, the most prominent of which was the shift of focus in the way in which the role and structure of language were viewed.

The concept of ESP should be placed and understood within the broader context of foreign language teaching in general. As Brumfit (in Robinson 1980:10) suggests, it should be viewed as a 'new emphasis in teaching' rather than as an independent or separate field in its own right. It is also equally important to conceive of ESP as "developing from, and contributing to, the language teaching profession as a whole" (McDonough 1984: preface).

1.1.2 The Constituent Notions of ESP

Basically, there are three constituents in the notion of ESP. These are simply: language, English in this case, specificity, and purposefulness. This should not imply, however, that these notions are separable and can be treated in isolation from each other. On the contrary, they are essentially interrelated and one cannot be dealt with without reference to the other. For the sake of convenience I shall consider them under separate headings.

1.1.2.1 English (Language)

What language is and how it works have always been crucial factors in the practice of language teaching. In other words, the way in which a language was viewed and described has always resulted in views on how it should be taught. Although the language teacher should not necessarily base his/her pedagogic decisions solely on whatever description the linguist provides, s/he should make use of the wide range of insights that linguistics may offer. However, it is wrong to assume that whatever the linguist describes the language teacher 'applies'; language teaching and linguistics, though interrelated, are different fields of concern (cf. Wilkins 1972:216). Language teaching is essentially pragmatic which implies that as language teachers we should be drawing our principles from every relevant field of knowledge and not only from linguistics. Nevertheless, knowledge of language descriptions is important for the language teacher as it makes him/her more aware of, and more competent at, what s/he is doing.

In its early years, ESP fell under the influence of the structural approach to language description (McDonough 1984:53). Thus, the E in the ESP was viewed as a formal system of structures which is rule-governed. In the light of such a view, descriptions of language were produced in terms of discrete structures and lexical items which were converted into teaching materials. On the assumption that the language of science and technology was different from general everyday language, attention was given, in early ESP, to the description of the 'special' linguistic features of technical texts. This assumption gave rise to conceiving of ESP as 'English for Special Purposes'. The word 'special' was used in association with 'special languages' or 'registers' such as the language of medicine or the language of law. In this sense, the S in the concept of ESP was a reference to the description of the language rather than to the purpose for which it was taught/learnt. The term English for Special Purposes was later replaced by the term English for Specific Purposes on the grounds that the former could be confusing (Mackay and Mountford 1978:4).

'Special languages' or 'registers' were analysed using the criteria of (a) frequency of occurrence and (b) presence/absence of particular syntactic and/or certain lexical items. The results usually found their way into teaching materials. However, the quantitative approach was later rejected because the descriptions based on it were confined to the sentence level and therefore failed to account for the fact that language was to be viewed as discourse in communicative contexts (Widdowson 1979:55).

During the 1970's and due to the influence of Speech Act Theory, the English of science and technology was viewed in terms of rhetorical features. The major development was the shift of focus from viewing language as 'usage', i.e., formal categories of rules, into viewing it as 'use' i.e., functional properties in social contexts (ibid. 1979:8). ESP, in turn, had to accommodate the communicative approach and since then became identified with it. Mackay and Mountford (1978:19) argue as follows:

> "A view of language as communication has had a profound influence on the direction of ESP over the last few years. Both the language to be taught and the purpose of teaching it have, as it were, came into focus more closely as a consequence of the notion."

Under the communicative approach, the level of language description was extended beyond the sentence and became discourse-based. Terms such as 'notions', 'functions', 'acts', 'moves' etc. became the bases on which syllabuses and teaching materials were built. (cf. Wilkins 1976, McDonough 1984:57.) Functions and notions are often used interchangeably, although they are not necessarily identical. The relationships between particular notions/functions and their grammatical and lexical realisations were the subject of various research investigations (cf. Trimble 1985). However, the fact that functions/notions can be correlated to their linguistic exponents should by no means imply a one-to-one correspondence between the two. The same function/notion can be linguistically realised by several different syntactic patterns and lexical items. Similarly, the same grammatical pattern or lexical item may indicate more than one function/notion depending on the context (Widdowson 1979:15).

However, even the functional/notional approach to language description did not escape criticism, on the grounds that the specification of 'illocutionary acts' was not enough and that larger stretches of discourse had to be taken into account. The subject of discourse analysis is much in vogue in ESP but the terminology employed can be confusing. There are many labels and, as suggested by Widdowson (1979:257) "it is not easy to find oneself around, let alone establish significant principles of investigation". Although communicative language teaching is quite a popular concept among language teachers nowadays, I think there is a long way to go before an adequate theory of describing the E in ESP will emerge. This, however, does not imply that ESP practitioners have to wait until linguists provide such a theory. After all, ESP is a practical business.

1.1.2.2 <u>Specificity</u>

Several factors have contributed to the increasing specification in language teaching. These are:

1. A widely felt dissatisfaction among foreign language teachers and learners about the content of materials and

the ways in which such materials were presented. The foreign language was, in many cases, the subject matter as well as the purpose of the teaching/ learning process. In addition to the absence of specified goals, the learner's perspective was much neglected; the aims, interests, wants and needs of the learners were hardly accommodated in language teaching programmes during the pre-ESP era.

2. Changes of 'geo-linguistic' nature. According to Strevens (1980:105), the role of English became to be seen as a means to achieve clearly defined practical goals in certain restricted areas. This implied the need to specify exactly what was needed. In various newly independent countries, the language policies were more selective in 'licensing' the areas of language teaching required.
3. Changes within linguistics itself, i.e., the tendency to specify the language teaching goals together with the tendency among linguists to view language as communication in context (Candlin, in Mackay and Mountford 1978:vi).

I have also mentioned earlier that the S in ESP was first referred to as _special_ and then later became _specific_. The term 'special' is misleading as it may imply that ESP is no more than the teaching of a 'special' language or a 'register'. In other words, the term 'special' is easily associated with the notion of a 'special language' that has special features making it different from the everyday language.

On the other hand, the term 'specific' has a different implication. It is a reference to the purposes which have been made no longer general or vague; it is a reference to the goals _for_ which the language is taught/learnt and not to the special features _of_ the language which is taught/learnt. This distinction is significant as the implications differ. Specifying the aims determines not only the precise language teaching content needed, but also the skills and the range of contexts in which the language is to be used.

Specificity, of course, implies the notion of non-generality. As mentioned earlier, ESP is not to be seen as a separate field from ELT. What makes ESP distinct from ELT, however, is the degree to which the purposes in teaching/learning the language have been specified. In general language teaching, as, for example, in school English courses, it is often the case that the purposes of language teaching/learning are either absent or are only vaguely stated. In ESP, on the other hand, the designer, the teacher, the sponsor and the learner are all supposed to be working towards the achievement of clearly identified and well-defined objectives. Robinson (1980:6) differentiates ESP from general ELT and states that

> "... the general with which we are contrasting the specific of ESP is that of general, education-for-life, culture and literature orientated language course, in which language itself is the subject matter and the purpose of the course."
>
> (original emphasis)

The theoretical implications of specificity are discussed by Widdowson (1983:5) where he compares specificity with generality in language teaching and argues that there exists no such difference between ESP and GPE (General Purpose English). Both are specific and both are purposeful; if there is a distinction, it lies, Widdowson argues, in the manner in which purpose is defined and implemented. Thus, specificity as well as generality of language teaching seem to be determined by the way in which the notion of 'purpose' is understood. In sum, specificity is central to ESP but should not be exaggerated. Much of what applies to general ELT is applicable to ESP. In ESP, specificity can be applied to various levels. This includes specifying purposes for which the language is taught/learnt, e.g., academic purposes, professional purposes, the subject matter of the language taught/learnt, e.g., Medical English, Legal English, the communicative contexts in which the language will be put to use, e.g., hospital clinical training, and the communicative needs of the learner, e.g., report writing.

1.1.2.3 Purposefulness

The notion of purpose is related to the notion of learning the language in question. In ESP, we are concerned with what the E is learnt for as well as with what the E is like. In traditional language teaching, the language itself was viewed as the purpose as well as the medium of language learning (Robinson 1980:6). The idea of purposefulness was rather secondary, if not missing, and the language was taught as a subject matter for its own sake. This was particularly apparent at primary and secondary school levels where learners, and to some extent teachers, of the foreign language were not clear about why they were doing what they were doing and only seemed to be aware of preparing themselves for public examinations. The foreign language was taught with no immediate or specific target in mind; it was to be taught and learnt within the school curriculum simply because it was there. Such a view of purpose would not suffice when we consider language learning at the tertiary level of education. Students at university and college institutions are often under time pressure and more aware than primary- and secondary-level students of why they need to attend a course in a foreign language.

Purposes in ESP are often identified with the sort of social contexts in which the language is or will be used. Such contexts are primarily to do with study and/or work situations. Widdowson (1983:6) defines the notion of purpose as "the eventual practical use to which the language will be put in achieving occupational and academic aims". The manner in which purpose is often indicated and categorised in ESP is in terms of the subject matter studied or the profession in which the learner is involved. However, it is observed that ESP has suffered from too much labelling in specifying purpose. This is not healthy for ESP as it may leave the impression that such sub-branches of ESP, e.g., Business English, English for Science, English for Hotel Staff, Technical English are actually quite different from one another (McDonough 1984:preface).

Various taxonomies of purpose are available in the ESP literature (e.g., Widdowson 1983:9, Robinson 1980:7, McDonough 1984:6, and Strevens 1980:110). They all overlap, but broadly speaking, purposes can be summarised under four headings (cf. Mackay and Mountford:2):

1. academic (e.g., English for students of medicine)
2. professional (e.g., English for doctors, lawyers)
3. vocational (e.g., English for hotel staff)
4. occupational (e.g., English for airline pilots).

Widdowson (1983:9) is strongly critical of the way in which the notion of purpose has been handled in ESP. "The kinds of ESP that have been suggested make up an observational list and have no status in theory", he argues. In addition, and apparently due to a lack of a sound theory of ESP, the relationship between the various branches of ESP, according to Widdowson, is to date far from being clear. He raises the question of the way in which the 'type of purpose' can be defined, and strongly objects to the interpretation of purpose in terms of "simply what people need to do with their language" in association with particular academic disciplines or professional activities. Rather, he proposes that the purposes in ESP should be "arranged along a scale of specificity with training at one end and education at the other". The rationale for such a combination of both educational and training purposes is Widdowson's belief that there is no difference in specificity and purposefulness between ESP and GPE (General Purpose English). GPE is seen as purposeful but the purpose is viewed as "a formulation of objectives which will achieve a potential for later practical use". Purpose is there but a delayed one in the case of GPE. Widdowson also stands against the view of purpose in ESP as "essentially a training exercise". He objects to the notion of purpose as a way dealing with a limited range of anticipated problems as in the case of air-traffic controllers, i.e., training.

Implicit in the notion of purpose is the notion of competence development, often employed when stating the objectives of a given language course. In this respect, it should be noted that the development of competence as the purpose in ESP does no longer refer to only linguistic competence in Chomskyan terms. In ESP, we are not concerned merely with the knowledge of the native speaker about the formal aspects of the system of his/her language. As a matter of fact, the notion of competence has been extended to incorporate not only the knowledge of the language system, i.e., rules of usage, but also the knowledge of the social rules that determine the appropriate use of the linguistic forms (Widdowson 1983:7). This is what is commonly referred to as the communicative competence as opposed to the linguistic competence of the learner. Thus, taking this view into account, the purpose in the teaching of ESP is no longer viewed as merely the development of the learner's knowledge of grammatical patterns and vocabulary items, but also of the rhetorical functions and the discoursal patterns of the language.

The shift from viewing the purpose of language learning/teaching as linguistic competence into perceiving purpose as communicative competence has come as a result of the changes that were taking place in linguistics in the 1970's. Hymes (1972:278), for instance, has stressed the importance of communicative competence and referred to the language user's ability to adapt the structures and meanings of the language to the social context in which language is employed. Communicative competence is an indication of the language user's ability to use the appropriate linguistic items, lexical and syntactical, in the appropriate social circumstances with the appropriate persons for the appropriate aims. The ability to communicate has become "an essential element in the definition of any ESP course" (Robinson 1980:11). This implies that in designing any ESP syllabus, the objectives or purposes to be achieved should be formulated in behavioural and functional terms.

No matter how we may conceive of the notion of purpose in ESP, the important point is that purposes or objectives have to be stated clearly and in as measurable terms as possible. It is important to be able to evaluate the extent to which the learner has succeeded in attaining the specific purposes of the course.

The criteria for assessing the achievement of the objectives should by no means be confined to passing examination papers , but rather be extended to include the degree to which the learner has been successful in actually using the language in the particular academic or professional contexts in which s/he needs to use the language. Passing a multiple-choice test on the knowledge of certain grammatical and/or lexical items is no guarantee that the learner has actually achieved the real specific purposes for which s/he is learning the language.

To sum up, ESP involves the following principles:

1. ESP is a pragmatic business; what is best is what works best.
2. ESP has not yet developed a theory of its own.
3. ESP is essentially learner- and learning-centred.
4. ESP is a reorientation of focus in ELT rather than a separate field on its own.
5. ESP lays emphasis on the communicative or functional aspects of language, but does not ignore the linguistic aspects of it.
6. ESP is a revolution against redundancy, irrelevance and inappropriateness of language teaching materials.
7. ESP is time-sensitive and cost-effective.
8. ESP is interdisciplinary. It is interdependent on other fields of knowledge, e.g., linguistics, anthropology, psycholinguistics, sociolinguistics.
9. ESP is a service and not an end.
10. The S of ESP is a reference to purposes specified and not to special registers.
11. ESP is relatively young and many issues in it are still controversial and unresolved, e.g., vocabulary, authenticity, communicativeness, teacher-training.

1.2 Rationale for this Study

Since the 1960's, a great deal of activity and achievement has taken place under the umbrella term of ESP. The literature published, in various forms, is growing steadily and research into many aspects is being carried out professionally; university and college departments concerned with language teaching have already begun to accommodate ESP. In short, ESP has developed into an 'institution' and an 'impressive industry' (Widdowson 1983:9).

However, although the ESP approach is now widely recognised and often positively viewed, one can observe that it is still far from being settled. The notion of ESP is often misinterpreted and a number of issues involved or implied by the notion remain unclear. There are still a number of gaps in our understanding of the subject and such gaps can only be filled by further research. Indeed, "ESP is still subject of much discussion and many key issues remain unresolved" (Robinson 1980:1). One such key issue which is still very much under-researched is the theme of this book, i.e., the ESP learner-dictionary interaction and the role of pedagogical lexicography in furthering ESP learners' lexical proficiency.

The research reported on here is basically an enquiry into aspects of the relationship between a group of ESP learners, who are student nurses at the University of Jordan, and the dictionaries they use. Considerations of practical and theoretical nature have motivated me to undertake this study. My ultimate aim is to contribute to the present knowledge of ESP by investigating an area that has received little attention in the literature from either ESP practitioners or lexicographers, i.e., ESP learner-dictionary interaction and the role of the dictionary in an ESP programme. My immediate concern is with the teaching situation in which I am currently involved, i.e., the ESP/EFL programmes at the University of Jordan.

At the theoretical level, it is my aim to assess the value of dictionaries in the teaching of ESP and from an ESP perspective. It must be stressed here that this study is carried out by an ESP practitioner and not by a lexicographer. This is not to imply that ESP and lexicography are not interrelated. Indeed they are, but the point is rather that this research is orientated towards, and motivated by, the needs and problems of the ESP learner as user/consumer of lexicographical products. My position in this respect is not very different from that expressed by Opitz (1979:89):

> "I am not a lexicographer. I am a teacher. My interest in lexicography derives from the fact that my students must - in varying degrees, consult dictionaries to do their work... It is only natural that I should be asked what dictionary I recommend, though I am rather afraid of that question."

In my case, throughout the various stages of my involvement in the learning and teaching of EFL and ESP, as a pupil being taught English as a foreign language at school, as a university undergraduate majoring in English, as a teacher trainee and as an in-service ESP practitioner the dictionary input, I felt, was too limited. I have not received instruction, formal or informal, on dictionaries and their use and other reference skills, despite the fact that it was very difficult for me to survive linguistically without a dictionary. As an ESP practitioner, I soon came to realise that in our ESP activities at the University of Jordan, the lexicographical input had been very insufficiently treated. The status of dictionaries was never made clear and in our designs, materials and tests, the dictionary question never surfaced. I believe that in our teaching situation we know very little about dictionaries and the relationship between our students and their dictionaries; and that the issue of reference skills input into our teaching has been left almost untouched.

With this in mind, I believe we have first to recognise that our knowledge of our students' reference skills and their interaction with dictionaries is quite limited; and second, without adequate knowledge as far as this issue is concerned it would be very difficult for us to claim that our ESP courses are successful and efficient. Simply, we cannot base our pedagogical decisions on sheer assumptions and/or impressions. A further motivation for this study is that in the wider context of English teaching in Jordan as a whole, to my knowledge, there has never been any research on dictionary use and/or reference skills teaching at any level in the Jordanian educational system. The situation can to some extent be generalised to the entire context of the Arab World, which is a large market for both ESP and lexicographical products. Of the rather slim body of literature on dictionary use, most of the studies reported have been carried out outside the Arab World and are therefore hardly applicable to Arab, especially ESP, learning contexts.

On the international front, both ESP and lexicography seem to lack detailed studies on aspects of dictionary use and the acquisition of reference skills among ESP learners/users. As far as ESP is concerned, a considerable amount of research has been reported on ESP-related issues, but the issue of ESP learner-dictionary interaction within the broader context of vocabulary learning/teaching has received scant attention from ESP practitioners. On the other hand, the published literature on lexicography seems to be either lexicographer-oriented rather than user-oriented and/or biased towards learners of English in general rather than learners of English for specific purposes.

The Aston University ESP Reference Collection is, for example, one of the biggest archives of ESP literature, both published and unpublished. Yet, a look at the 1987 catalogue of this collection shows that studies on dictionaries and their uses/users and/or reference skills acquisition among ESP learners are quite few in comparison to the relatively rich body of literature on, for instance, the use of video in language teaching or on language description of ESP varieties. A similar look at the content pages of the ESP Journal, since its birth in 1980, shows how scant is the attention to dictionaries and their use by ESP students.

What initially inspired the idea of this research was a key article by a leading ESP practitioner entitled "Vocabulary Work in LSP: a case of neglect?" (Swales 1983). Swales argues that the lexical problem is in fact everybody's problem: the learner, the teacher and the lexicographer. He notes that

> "It is often said that the day-to-day problems of LSP learners are frequently lexical. I can certainly assert that the day-to-day problems of LSP teachers are also frequently lexical, and I would be surprised if that were not also the case for specialized lexicographers and lexicologists." (ibid.:32)

In this context, Swales laments the serious neglect of the dictionary user's interests within the European tradition of LSP work. The situation is even worse when we consider the Anglo-American and the Third World tradition of LSP, he argues; here, not only have the user's interests been neglected, but the lexical level itself has also been ignored (ibid.:22).

In the last two paragraphs of his recent book, Swales (1985b:213-214) confirms that "the situation still remains very unsatisfactory". He argues that despite the fact that there have been a number of attempts to apply collocational and componential analyses to ESP, the outcomes were rather 'uncertain'. For example, "few students seem able to cope with the complexity" of the LONGMAN DICTIONARY OF SCIENTIFIC USAGE by Godman and Payne (1979) although it is intellectually 'impressive'. Swales concludes his book with the following observation:

> "Unfortunately, it is precisely that interest in process and interaction which is missing in vocabulary work; interaction between specialized vocabulary support materials (technical dictionaries, glossaries, lexical fields, etc.) and the language learner and user are under-researched, and the process of technical and subtechnical vocabulary acquisition imagined rather than investigated."
>
> (Original emphasis)

This clearly indicates that whereas interest was placed around the production of the lexicographical work, the user/customer was left to his/her own devices in coping with the complex task of vocabulary learning. Moreover, despite the fact that the vocabulary problem is the concern of everyone in both the ESP and the lexicography processes, there seems to be a gap between the two fields. For instance, it is rare to find a lexicographical product that has been based on research results gained from what actually goes on in the teaching and learning of ESP; and it is also hard to find evidence of ESP practitioners who have been involved in the production of lexicographical works.

Another EST (English for Science and Technology) authority has recently made an observation which is, in my opinion, worth considering:

> "For some unexplained reason, few non-native science and technology students seem to have specialized dictionaries or, in many cases, are even aware of their existence."
>
> (Trimble 1985:129)

Though the author does not elaborate on this point at any length, his note deserves attention. One could ask: why should this happen? and what is the 'unexplained reason' that leads non-native ESP learners not to acquire technical dictionaries? I suspect that in many ESP contexts, including mine, many students, and probably teachers, are simply unaware of lexicography.

Certainly, dictionaries have a role to play in the acquisition and development of ESP vocabulary; but this issue cannot be taken for granted. According to Kennedy and Bolitho (1984:56) "it is dangerous - perhaps more so in ESP than in 'general' English - to assume that vocabulary will take care of itself, or that students will absorb words by 'osmosis'". At least in my own teaching situation, such assumptions are still being made. Yet, few of my students would not consider vocabulary as a major obstacle in learning the subject matters of their courses (cf. Meara 1980:221).

The teaching of vocabulary, having recovered from years of neglect, is currently the subject of discussion by researchers and practitioners who are currently looking into possible ways of improving the situation. Obviously, among such ways is the potential role of the learner's dictionary. Pedagogical lexicography, having begun to "flourish as an academic discipline" (Hartmann 1986a:1), and become recognised as a profession, has taken on a significant role in the teaching of English as a foreign language. "One important impetus to vocabulary learning may be coming from the field of lexicography", Hartmann (ibid.:7) argues. Few would dispute this observation as it is becoming obvious that the isolationist era in the relationship between lexicography and English language teaching is moving towards its end. Not only can ESP benefit from lexicography, but it is also important to assess the potential of ESP in lexicography. Hartmann (1981a:299) argues:

> "One of the most fruitful notions in applied linguistics in recent years has been that of 'purpose specificity': in EFL teaching, in technical register rhetoric, in testing and in translation studies the seed of purpose-specific problem-solving has taken root. The time has come to try this approach in lexicography."

The revived interest in the notion of the learner's dictionary is certainly a trend to be welcomed, as it is a well-known fact that dictionaries do contain a lot of information which can be useful to language learners. However, it must be noted that little attention has been paid so far to dictionary uses and/or its users despite the fact that "the dictionary is a recognised institution" and "the most successful and significant book about language" (Ilson 1985:5 and 1). It is true that dictionary-makers, publishers and language teachers have recently become more interested to know about the needs, problems, and interests of the language learner/user; yet it is still being reported that "the dictionary as a learning tool has received relatively scant attention from the language teaching profession" and that "the function and use of dictionaries by learners and teachers has had occasional mention in the literature" (MacFarquhar and Richards 1983:111).

Traditional lexicography seems to have concerned itself with the problems of the lexicographer rather than the needs of the dictionary user. In the ESP tradition, we are told that "there has grown up a belief that by simply <u>producing</u> an LSP reference work most of the vocabulary problems of that LSP group of learners can be solved" (Swales 1983:21, original emphasis). Hence, there seems to be an obvious need for empirical research on who uses what dictionaries for what purposes in which contexts and with what degree of success. It is simply not enough to respond to our students' lexical queries by uttering statements like "Use your dictionary" or "Look it up" and assume that this is the end of the story.

In the literature of both ESP and lexicography, little is reported about the uses of dictionaries by ESP learners. Even the relatively few studies available have been subject to criticism. For example, Hartmann (1986a:18) notes that

> "Studies that exist are rarely representative, comparable or replicable: statistical and experimental techniques have hardly been tried. Tests are cumbersome, interviews are costly, questionnaires are biased, self-observation is of limited generalisability."

The message is clear. More and better research is needed and a variety of techniques have to be devised and tested if a better general picture of the position of dictionaries is to be gained.

Dictionaries for language learners are becoming increasingly sophisticated in content and format, whereas little is known about the learners' reference skills which are often described as rudimentary (cf. Cowie 1981a:206). In order to narrow the gap between ESP learners and the realm of dictionaries, this study attempts to contribute to a greater understanding of one of the key institutions in our profession, i.e., the dictionary which seems to have been taken for granted in ESP, which itself is an institution that is still not without deficiencies (cf. Brumfit 1985:v, and Widdowson 1983:1). After all, as Swales (1985b:214) puts it,

> "...in many ways English for Specific Purposes has grown to a considerable degree of maturity... but until we can find ways of ensuring the collaboration of learners, practitioners and specialized lexicographers and lexicologists in the experiment and trialling of new and more efficient ways of coping with the vocabulary problem, I do not think ESP can really be said to have come of age."

This study is a contribution to this collaborative approach called for by Swales.

1.3 Scope and Specific Objectives of this Study

This study has two dimensions. One is theoretical where it is intended to relate pedagogical lexicography to the context of ESP and where lexicography is viewed from an ESP point of view. An ESP approach to pedagogical lexicography is explored in the light of the results of this study. The other dimension is empirical where aspects of dictionary use in my ESP teaching situation are investigated. This has three advantages: (1) it provides the ESP practitioner in other situations with a comparable case; (2) it supplies the pedagogical lexicographer with insights from within a real ESP situation where modifications may be introduced in lexicography for ESP learners; and (3) it offers a database to the University of Jordan, and particularly the Faculty of Nursing and the Language Centre, where the results and the recommendations can be of direct use in future planning and the improvement of current ESP courses.

This is the first and the only study of dictionary use with particular reference to ESP learners at university level in Jordan and it is also probably the only one in the Arab World. Hence, there are no available ESP-oriented studies with which to compare it. However, an attempt is made throughout this study to compare the results with other available results of dictionary-use studies of general users, to find out whether or not such studies can be applicable to ESP groups of learners.

One important aim of this study is to encourage other researchers in the field of ESP to follow up this work with similar studies but on different groups of ESP learners in different situations and possibly using different techniques. This can pave the way for a new trend in ESP research, i.e., contrastive user-research.

In particular, this work aims to answer the following questions:

1. Do the ESP students come to the ESP programme with adequate dictionary reference knowledge and skills? How much knowledge and skill can they be assumed to have on joining the ESP programme? Have they received adequate instruction on dictionary use at pre-university levels? What is the status of dictionaries in the curricula of English at the Compulsory and Secondary school levels in Jordan?
2. What is the status of dictionaries in the curricula of English at the University of Jordan's ESP/EFL programmes?
3. What are the attitudes of the ESP students, ESP staff and subject (nursing) staff to dictionaries and their use?

4. What strategies do the ESP students employ in dealing with dictionaries?
5. What are the communicative needs of the ESP students as indicated by students and ESP as well as subject staff?
6. What are the communicative contexts in which dictionaries are being used in relation to communicative needs?
7. For what types of information do the ESP students use dictionaries in relation to communicative needs?
8. Which and what type/s of dictionaries do the ESP students use?
9. Are the dictionaries mostly used by the ESP students appropriate and capable of meeting the students' needs?
10. Do aspects of dictionary use vary according to the students' level of study, i.e., first year, second year etc.?

CHAPTER TWO

DICTIONARY-USER INTERACTION: A REVIEW OF THE RELEVANT LITERATURE

2.1 Introduction

The user-perspective tradition of research in lexicography is relatively young. It goes back to 1960 when the Conference in Lexicography was held at Indiana University, U.S.A. The first, and perhaps the most significant, recommendation the conference unanimously agreed upon was that

> "Dictionaries should be designed with a special set of users in mind and for their specific needs, e.g., an English-Arabic dictionary for American users for help in speaking Arabic, ..."
>
> (Householder and Saporta 1962:279)

This recommendation, interestingly, came at a time when ESP first came into existence as a professional movement in language teaching. Swales (1985b:x) points out that ESP "has no clear and indisputable beginnings", but chooses 1962 as the beginning of the ESP story in which Barber's article "Some Measurable Characteristics of Modern Scientific Prose" serves as the first episode. In what follows I shall review, chronologically, the major empirical studies that have been reported in the literature of dictionary use.

2.2 Barnhart (1962)

The first pioneering contribution to the user-perspective tradition of lexicographical research was that of Barnhart (1962:161) who emphasised that the role of the dictionary is to "answer the questions that the user of the dictionary asks" and that "dictionaries on the commercial market will be successful in proportion to the extent to which they answer these questions of the buyer". Barnhart's study was conducted in 1955 and involved 108 questionnaires which were sent to teachers of English writing in American high schools in 27 States. The purpose was to find out the relative importance of different types of information entered in popular American college dictionaries. The teachers were requested to rate six types of information according to the importance attached to them by freshmen students. The results were reported to have shown that the most frequently looked-up information was meaning, followed by spelling and pronunciation. Synonyms and usage came next and etymological information came last in priority. The study also revealed that the inability to recognise the root form was one of the major spelling difficulties with derivatives.

In his critical review of user studies, Hartmann (1987:13) points out that Barnhart's study was based on a relatively small sample and that the questionnaires and the numerical results were not reproduced in Barnhart's paper. A further criticism was the fact that his results were based on indirect responses by teachers rather than on direct observations of the users surveyed. Nonetheless, such criticisms should by no means be taken to underestimate the valuable contribution of Barnhart. It was the earliest attempt to provide quantitative data on functions of the dictionary taking the user-perspective as an approach. The real value of Barnhart's survey was its bringing into sharp focus a new and a very significant criterion to lexicography viz., the user-perspective.

Contrasting Barnhart's study to my research, I should like to mention here that Barnhart undertook his survey more than three decades ago (1955) in an American setting on dictionary use by native speakers of English, whereas I am concerned with an ESP programme where Jordanian student nurses study English as a means of learning their specialist subjects in the late 1980's. However, Barnhart's study remains valuable as a case for comparison. In fact, I learned from the criticisms directed against Barnhart's study that my group of subjects should be of a larger size and that responses should be elicited directly from the learners themselves, though considering the teachers' responses would be of additional value.

2.3 Quirk (1973)

The second episode in the story of user-based lexicographical research was the contribution made by Quirk (1973). In the academic year 1971-72, Quirk examined the use of dictionaries by 220 undergraduate British students at the University College, London. Represented in the sample were both sexes and science as well as arts students. The research tool was a questionnaire which consisted of 30 questions on various aspects of dictionary use such as ownership of dictionaries, frequency and purpose of dictionary use and reported problems in dictionary look-up.

The study revealed that dictionaries were owned by the majority of the subjects. It was also found that these users referred to their dictionaries regularly (71% used dictionaries at least once a month). As for the types of information looked up, Quirk found that meaning was the main purpose of dictionary use and that the subjects expressed interest in synonyms and antonyms. Unfavoured information in dictionaries was pronunciation, grammar and etymology. In addition, arts students were said to have made greater use of dictionaries than science students. The lack of interest in pronunciation in dictionaries was attributed by Quirk to the users' attitudes, i.e., indifference rather than difficulty in understanding the phonetic system in dictionaries. The subjects were also reported to have experienced difficulties with understanding the metalanguage of definitions.

Quirk's work was pioneering as far as the use of dictionaries by British students was concerned. It added new dimensions such as the exploration of the users' attitudes and preferences. For example, the subjects were asked to suggest improvements on the various aspects of dictionaries in use. Among such suggestions were that definitions should be made more intelligible, and that improvements could be made as to the layout of dictionaries because they felt the dictionaries had too many entries which were difficult to follow.

In comparison to Barnhart's study, Quirk's work made a real advance in the user-perspective approach. Though of great value methodologically, Quirk's contribution remained bound to its context, i.e., British university students who were native speakers of English using monolingual English dictionaries. It did not tell us about the use of bilingual dictionaries or the learning of ESP by non-native speakers of English. Nonetheless, I found Quirk's results interesting; for example, the arts versus science students tendencies in the use of dictionaries. Also of interest were his subjects' complaints about intelligibility of dictionary entries and dictionary layout. If that was the case with native speakers of English, I suspect that non-native ESP learners would express similar complaints.

2.4 Bujas (1975)

This study aimed at assessing the appropriateness of a particular dictionary to the needs of a particular group of users. The dictionary in this case was bilingual medium-size general-purpose English-Croatian; the subjects were 18 undergraduates at the University of Zagreb in the Department of English. Bujas wanted to test the performance of the dictionary against the requirement of reading British and American topical texts drawn from sources such as Time magazine and The Observer.

Over a period of more than two years, the subjects were requested to skim through a total of 34 issues of 9 periodicals with the purpose of finding instances of inadequacies in the dictionary under study. The methodology involved dividing the subjects into two groups: one looked for missing entries only whereas the other had the task of looking up every word in the texts. The study showed that the subjects did spot inadequacies in the dictionary in terms of missing and/or translation equivalents and absent collocational entries. The subjects also managed to compile a list of 6,272 vocabulary items of which 78.3% were recommended for inclusion in later versions of the dictionary (cf. Hartmann 1987:18).

Bujas's study was a limited one; it involved a small sample of students reading English as a subject rather than as a means of understanding and expression in a specialist field such as Medicine or Nursing. It was also based on a single bilingual dictionary involving English and a Slavonic language. Nevertheless, the methodology employed by Bujas was original, and according to Hartmann (ibid.), "resulted in a workable procedure for improving an existing bilingual dictionary".

2.5 Tomaszczyk (1979)

Motivated by the belief that among learners of English as a foreign language there had been a widespread dissatisfaction with dictionaries while little was known about the position of dictionaries in the language learning process, Tomaszczyk undertook his informative investigation. The context of his study was Poland and the U.S.A., and the sample consisted of almost 450 subjects who were Polish and foreign learners of English as a foreign language at Polish universities and Polish as well as American language teachers and translators .

Tomaszczyk's results indicated that bilingual dictionaries were used almost exclusively by the subjects, and that L2-L1 dictionaries were more appreciated by the subjects than the L1-L2 dictionaries. Monolingual L2 dictionaries were, however, found to be of secondary importance to the users. As far as the contexts of dictionary use were concerned, the study reported that writing and reading came first, translation next, and listening and speaking last in terms of priority. Interestingly, not only were less advanced EFL students reliant on bilingual dictionaries, but also the supposedly advanced users, such as university language teachers. The study also showed that awareness of the content of dictionaries was not high among many of the beginning and intermediate student subjects. There was evidence that the respondents, like those of Quirk's research, felt dissatisfied with their dictionaries. Tomaszczyk's subjects preferred encyclopaedic information, e.g., pictures and drawings, to be included in dictionaries. As for the types of information looked up, it was reported that meaning and spelling were the main concerns of the users.

No doubt, Tomaszczyk should be credited for his comprehensive research. However, mailed questionnaires might not always be reliable and therefore researchers into dictionary use using such instruments need to be as near as possible to the user and to complement questionnaires with follow up studies and/or consider varying their methods. In addition, Tomaszczyk's work said nothing about the needs of ESP students and was concerned with respondents of various levels and different needs in Poland and U.S.A.

2.6 <u>Opitz (1979)</u>

Of more relevance to ESP was the study by Opitz (1979). In fact, Opitz's work was the first, and one of the very few, investigations to look at the ESP learner user. Swales (1983:21) has described it as pioneering in the European tradition of lexical studies which "has tended to be insufficiently interested in interactions between reference works and their potential customers". Opitz (1979:89) stressed the user-perspective in the very first passage of his article. He spoke of himself as a teacher rather than a lexicographer and pointed out that his perspective was "that of reception and use, and not so much the production, of dictionaries."

The main objective of the study was to assess the appropriateness of particular specialised and general bilingual German-English and specialised and general monolingual dictionaries in relation to the needs of German nautical postgraduate students and "adults from various Engineering professions in courses of further education" at Hamburg Polytechnic. He noted that lexicography was heading towards specificity and raised a number of valid questions to which he attempted to provide "at least a partial answer". The questions (reproduced here in a slightly modified format) are worth considering as they touch upon a very important, and often neglected, sphere of lexicography:

1. Is the trend towards specialisation in lexicography necessary and useful?
2. Why is it so difficult to find a truly polytechnical dictionary in one volume?
3. Would it not be worthwhile to abstract that common lexical core from the various specialised registers and cast it into the form of a truly basic lexicon?
4. What are the criteria of lexicography?
5. Who are dictionaries, and specialised dictionaries in particular, meant to serve? The specialist? The layman? The learner?
6. Who really needs a dictionary, and what kind?
7. Which parts of the dictionary are really made use of, and what percentage of its substance is more ballast, the stuff that justifies the hard covers?
8. Do dictionaries serve any real purpose if they double up as encyclopaedias?
9. Should dictionaries adopt a diachronic perspective?
10. Should dictionaries be monolingual or bilingual?

Motivated by such questions and made curious by his students' remarks on inadequacies in bilingual English-German dictionaries when applied to nautical texts, Opitz selected a 1000-word text from a nautical periodical and asked his students (number not cited) to identify the words they thought related to 'nautical affairs'. This first part of the study showed a "wide divergence in frequency of the items thus defined". The same text was then given to a non-nautical group of engineers to determine to what extent 'semantic categorising' was affected by subject familiarity. The results showed that both groups differed in their assessment of the technicality of the items concerned.

The experiment was taken further to involve checking whether or not, and if so, how the identified items were entered in various dictionaries used by the subjects. The results showed a significant variation among the dictionaries considered in terms of number of specialised entries included and in terms of definition and labelling styles. Opitz concluded that:

1. A specialist dictionary would best serve its intended purpose if it fully espoused the fact of its specialisation and did not attempt to follow other models. This would apply both to the lexical corpus and the format used.
2. The student users of a specialist dictionary expected efficiency in the dictionary they used, i.e., they did not need material that was seldom demanded: they needed specific answers to specific questions fast and with minimum linguistic complexity.
3. The compiler of a specialist dictionary should seek expert technical guidance.

The value of Opitz's contribution could hardly be disputed. It opened new horizons and invited researchers as well as practitioners in teaching and lexicography to consider the role of dictionaries, particularly technical ones, in the learning of specialist lexis by ESP students. However, the study was of a limited scope and generalisability. It was concerned with the German context where the subjects were postgraduate students learning English for professional purposes and told us nothing about undergraduate students studying English for academic purposes.

2.7 Baxter (1980)

In order to find out about the needs of his students, their dictionary habits and preferences, and to explore the influence of bilingual versus monolingual dictionaries, Baxter collected information from 342 subjects. These were Japanese students of English but of varying subject specialisations; 18.1% were majoring in English and came from faculties of Law, Letters, Humanities and Education and 18.9% were non-majors in English who came from faculties of Education, Economics, Agriculture, and Engineering. It was not clear, however, whether such students, majors and non-majors, were following a specially designed ESP programme or not.

The data were collected through a questionnaire which was administered in late 1979 at 3 universities in Japan. It was not mentioned whether the questionnaire was mailed or otherwise administered and whether the language of the questionnaire was Japanese or English. The level of students was not equally represented in the sample. For example, almost 60% of the sample were second-year students while the fourth-year students represented only 7.6% of the total.

Baxter concluded that many other student populations could be assumed to exhibit similar patterns insofar as the predominance of the use of bilingual dictionaries was concerned. On the dictionary selection question, Baxter reported that his students preferred to resort to bilingual dictionaries because they were 'easier to use' in contrast with monolingual English dictionaries. He therefore put forward a strong argument in favour of a learner's monolingual English dictionary which employed controlled defining vocabulary.

However, Baxter suggested that students' feelings about bilingual dictionaries should be taken into account when introducing the use of learners' monolingual English dictionaries, as students would be expected to have developed certain learning strategies associated with the use of bilingual dictionaries and therefore could not be assumed to welcome the shift towards monolingual dictionaries. Hence came his plea to train students in using monolingual dictionaries in a systematic manner, where reference skills instruction was integrated into the teaching/learning process. I found Baxter's survey very insightful and therefore decided to incorporate some of its findings into my own study. For example, the question of dictionary choice would be of direct relevance to the teaching of ESP, and the argument on accommodating students' attitudes towards dictionaries and their usefulness was highly valid.

2.8 Béjoint (1981)

Béjoint (1981) was a modest, but certainly very informative, survey in the literature of user-oriented lexicography. Béjoint took up Tomaszczyk's remark (1979:103) that the needs of the popular commercial dictionary audience had been neglected, and explored "the virtually unknown territory of the users' reference skills and habits" (Béjoint 1981:208), The users in this case were 122 French students of English at the University of Lyon, France, of whom 63 were in their second year, 43 in their third year and 16 in their fourth year; most of these students were teacher trainees of English. A questionnaire of 21 questions, worded in French, was used in collecting the data. The questions touched upon important aspects of dictionary use such as dictionary ownership, dictionary typology, reasons for choosing particular dictionaries, preferences, frequency of dictionary use, information looked up, contexts of dictionary use, problems in dictionary use and attitudes towards dictionaries.

Béjoint found that the vast majority of the sample owned a general monolingual English dictionary, mainly British learner's ones, and that the teachers' recommendations were the main reason behind dictionary choice. Students seemed to prefer the dictionary they usually worked with and to value the dictionary that covers most items. The study also revealed that dictionaries were mostly used for decoding purposes and that meaning was placed by the majority of students among the three types of information most often looked up, followed by grammar, spelling, pronunciation and etymology.

The study also showed variations in dictionary use according to the age and study level of students. More than half the sample said they browsed through their dictionaries but a similar percentage said they did not use the codes that indicated how a word should be used. The students were found to be generally satisfied with their monolingual dictionaries, but it was reported that they found that monolingual dictionaries called for much linguistic sophistication on the part of the user. Complaints about monolingual dictionaries were said to include unsatisfactory definitions, missing vocabulary, unsatisfactory grammatical guidance, unintelligible coding systems and lack of phonetic information.

Béjoint's contribution should be credited as it brought into focus important issues of interest to both lexicographers and language teachers. It called for an approach in which dictionary awareness ought to be integrated in the design of language courses as it informed us that many students were not even aware of the riches of their dictionaries. It was also illuminating as it provided us with a detailed profile of dictionary habits, preferences and complaints among language learners. Although the sample was relatively small and the type of context and dictionary users in Béjoint's study were different from mine, I found the methodology adopted beneficial and the findings worth considering in comparison with my own. In particular, the questions on browsing and reference to dictionary appendices in my survey were, inter alia, insights from Béjoint.

2.9 Bensoussan et al. (1981)

Bensoussan, Sim and Weiss reported on two pieces of research on the relationship between dictionary use and students' performance on a reading comprehension test. Although conducted separately and independently from each other, at two different universities in Israel, both studies had a unified objective, i.e., to test the hypothesis that dictionary use has an impact on the achievement on an EFL reading comprehension test. The subjects of the experiment were students of a similar proficiency level in English, attending reading comprehension courses at Haifa and Ben Gurion Universities during the session 1977-1979. The students were assumed to have had seven years of school English and to have received instruction on how to use a monolingual English dictionary. At Haifa University, 700 first-year students attending the advanced reading programme were given nine reading tests. Each student received a 600-800-word text and 20 questions and was required to indicate the time spent on the test and the choice of dictionary used.

The study at Ben Gurion University involved 91 students with a similar level of proficiency to the ones at Haifa. The students were of different mother tongues, but the majority, 58%, spoke Hebrew as a native language. Three reading texts (each of 500-700 words) were given to students with 10 multiple-choice questions for each text. Students answered the questions on each of the three texts under different conditions: one text without a dictionary, another text with a bilingual dictionary and a third with a monolingual dictionary.

The two studies yielded similar results. They showed that there was a general preference for bilingual dictionaries and, very interestingly, that no significant correlation was found between dictionary use and test results. A follow-up study was carried out at Haifa University in an attempt to "clarify the test results and to understand the underlying attitudes and expectations of dictionary users" (Bensoussan et al. 1981:217). A questionnaire on dictionary use and preferences was answered by students of English, their teachers, and another near-native group of 13 third-year students of psychology. The results indicated that many students expected the dictionary to be useful in comprehension but felt dissatisfied when it failed to give the exact meaning of a word in a particular context.

The significance of the contribution by Bensoussan and her colleagues must not be underestimated despite the fact that the results have been received cautiously by some (cf. Hartmann 1987:27.). It introduced an experimental dimension to the study of dictionary use in addition to the popular questionnaire technique. It also raised an issue where little research has been done, i.e., the value of dictionary use in the context of examinations. In my study, I decided to explore this aspect by inserting certain items on the subjects' attitudes towards using dictionaries in both language and subject examinations.

2.10 Jain (1981)

A common problem among foreign learners of English is that of polysemy, i.e., the understanding of important relations among word meanings. Jain (1981:277) notes that in "L2 lexicography a proper sense of equivalence is often blurred for the learner by the fairly common but circular glossing technique of the one-word synonym". Particularly for productive use of the language, learners usually face difficulties in establishing the right use of the word even with the help of a dictionary.

Jain illustrated this problem by an exercise which he conducted with his students. They were asked to find out from three learners' monolingual English dictionaries the contrastive meaning properties of certain words so that they could decide for themselves whether the words were interchangeable or not. For example, "I don't want you to move from there. Would you keep/remain standing for half an hour?". Jain reported that the distinction between keep/remain remained 'mysterious' even after the students went through the three learners' dictionaries.

The main lesson that could be learned from Jain's report might be that the three learners' dictionaries (the OXFORD ADVANCED LEARNER'S DICTIONARY, the LONGMAN DICTIONARY OF CONTEMPORARY ENGLISH and CHAMBERS UNIVERSAL LEARNERS' DICTIONARY) failed to give the learners enough guidance as far as the problem of polysemy was concerned. This suggests that pedagogical lexicographers have to develop better ways in order to meet the various challenging problems of language learners while language teachers should familiarise their students with learners' monolingual English dictionaries.

2.11 Ard (1982)

This rather limited case study reported on some aspects of dictionary use in relation to a writing task performed by two learners of English, one Japanese and the other Arab. The purpose of the study was to determine the degree to which written composition could be influenced positively or negatively by the use of a bilingual dictionary. Ard analysed real instances of dictionary use which involved students' video-recorded recollections of their interaction with bilingual dictionaries, and instances of items looked up when writing in class.

The results indicated that the use of a bilingual dictionary could be seen to cause certain lexical errors; however, there were instances of errors which were not necessarily due to such a use but may have been attributed to interlingual lexical interference. These results could not be generalised because of the very limited scope of the study. However, the technique employed was valuable as it provided researchers into dictionary use with a comparatively reliable way of observing dictionary-user interaction in context.

2.12 Hartmann (1983)

The user-perspective approach to the study of lexicography owes much to Hartmann. He has repeatedly called for 'more and better research' into various aspects of dictionary use. This, in his belief, is the only way to replace the often dogmatic and speculative literature about the role of the dictionary; empirical research would solve many of the puzzling questions that face the lexicographer, the language teacher, the language learner, the translator, and the publisher. In ESP, Hartmann's calls for user-based research have been paralleled mainly by Swales (1985b:214).

The 1983 study reported by Hartmann aimed at studying the use of bilingual dictionaries among learners of German in South West England; in particular, the study had the purpose of finding about what types of dictionaries were used, in what contexts and for what purposes. The subjects of the study were a homogeneous group of 17 teachers and 118 students at more than 200 educational establishments covering secondary, further and higher education. One questionnaire was filled in by each teacher and two by each student group in the sample. The study revealed that translation topped the activities in which the bilingual dictionary was mostly used, followed by reading and writing. Furthermore, semantic and syntactic information were found to be looked up more often than phonetic and etymological information.

Hartmann also reported that the learners received inadequate training in the use of dictionaries. One of the conclusions he arrived at was that the role of the bilingual dictionary should not be underestimated and, at a time when learners' dictionaries were so far behind in languages other than English, it would seem unrealistic to ban or discourage the use of bilingual dictionaries in favour of monolingual dictionaries.

2.13 **Mitchell (1983)**

This study was part of a research project investigating the strategies of reading in Scottish secondary schools. The purpose of the project was to divide the skill of reading into sub-skills and to evaluate students' strategies in learning these skills. This, of course, involved eliciting data on strategies of information retrieval from dictionaries. One such strategy was the recognition of the most relevant piece of information in the dictionary and then using it to understand the text in question.

Hartmann (1987:15) pointed out that Mitchell's study confirmed that the task of finding the meaning of a word in a dictionary was an essential part of, and an extremely complex process in, dictionary use. This task involved the search for the appropriate headword, understanding the structure of the entry, identifying the relevant part of the definition, relating the appropriate sense to the context provided, and merging the word with the context of the source text.

2.14 MacFarquhar and Richards (1983)

Lexicographers and publishers produce and sell dictionaries, but do they know how intelligible learners find definitions in dictionaries? "To determine what learners find helpful or comprehensible it is only necessary to ask them." This is just what MacFarquhar and Richards attempted to do in their study. The approach to the study involved selecting three dictionaries for comparison: the LONGMAN DICTIONARY OF CONTEMPORARY ENGLISH (LDOCE); the OXFORD ADVANCED LEARNER'S DICTIONARY (OALD); and WEBSTER'S NEW WORLD DICTIONARY (WNWD). These dictionaries were selected because they incorporated different defining styles.

The subjects of the study were 180 foreign students (from Asia and the Pacific) attending intermediate to advanced courses in listening, reading and writing at the University of Hawaii English Language Institute. 60 words were selected and their definitions compared. The word sample was deliberately weighted to be representative of the kinds of items learners were reported to consult dictionaries for; it included idioms, culture-specific words, slang terms and compounds. The entries for each test word were photocopied and assembled on individual cards, and a space above each entry was provided for the subjects to indicate which definition they found the easiest to understand.

The results showed the superiority of LDOCE over the other two dictionaries as the clearest in terms of definition intelligibility; OALD came second and WNWD third in priority. The results also indicated that definitions from the dictionary intended for native speakers were judged to be the least comprehensible. Moreover, it was shown that there was no evidence that the gaps between the dictionaries narrowed as proficiency increased.

The conclusions could be of great value to both language teachers as well as pedagogical lexicographers. On the one hand, they confirmed that dictionaries for language learners should be different in scope, format, content and orientation from those intended for native speakers; and on the other hand, they lent support to the principle of controlled vocabulary in dictionaries for learners. The methodology adopted was tight and impressive as the researchers took great pains to ensure a sound design and a set of procedures for the experiment.

Furthermore, the study was original in that it attempted to investigate an important, and often neglected, area in dictionary use, i.e., definition intelligibility in dictionaries not through self-appointed teacher or lexicographer judges, but rather using real learner users. However, as the researchers themselves admitted, the study had its own limitations. Both the sample of learners and the sample of words and entries were small and therefore the conclusions remained tentative. Also, the study investigated the learners' perceptions rather than how helpful the definitions actually were.

2.15 Chagunda (1983)

A study similar to MacFarquhar and Richards', but of more relevance to the teaching of ESP is Chagunda's M.Sc. dissertation on "Interfaces Between Technical Dictionaries and EAP learners". Using specialist informants, Chagunda set out to compare how technical terms were treated in technical dictionaries. Two dictionaries were selected: the LONGMAN DICTIONARY OF SCIENTIFIC USAGE (LDOSU) by A. Godman and E. Payne, 1979 and the MODERN DICTIONARY OF ELECTRONICS (MDOE) by F. Rudolf, 1963.

Chagunda visited the informants individually and asked them which definitions, in the two dictionaries, appeared to be adequate to them and for what reasons. When an informant identified a definition as inadequate or improper s/he was asked how s/he would define the term in his/her own words. The respondents' reactions were tape-recorded and the transcripts were analysed. Chagunda reported some interesting results. For example, it was found that technical students generally disliked extravagant definitions, and that students appreciated the usefulness of diagrams in a technical dictionary.

As Chagunda described it, this study was a 'modest survey', but nevertheless, I think it is worthwhile. The use of specialist informants as a means of eliciting data on dictionary use should be encouraged as it might yield reliable information directly from users. ESP lexicographers should also learn from the finding that all students in the study agreed that illustrations were of crucial value in dictionaries as they would enable the user to visualise what a term entailed.

2.16 Greenbaum et al. (1984)

The study reported by Greenbaum, Meyer, and Taylor (1984) aimed at assessing the dictionary image among American college students, and comparing the findings with Quirk's survey (1973) of British students. A group of 240 arts and science students at the University of Wisconsin were chosen for the study. Among the variables investigated were the students' field of study, frequency of dictionary use, sex, and year of study. Significant differences were found between the arts and science students which might suggest that the type of subject matter studied could have an influence on the extent to which dictionaries were used.

The study also reported that American college students used dictionaries more frequently than British students. As for the type/s of information looked up, the American students searched mainly for meaning in their dictionaries but, unlike the British students, they seemed to have experienced more problems with spelling. The types of dictionaries used by the American students were said to be different from those reported by British students.

Greenbaum et al. have certainly contributed to the growing, but still fragmented, profile of dictionary uses and users. What I learned from this study, among other things, was that it was necessary in my survey to explore whether or not various aspects of dictionary use varied according to study level of the subjects. However, the study reported by Greenbaum et al. was on native speakers of American English and, though insightful, had little to say about the needs and problems of foreign learners of ESP.

2.17 Hatherall (1984)

It could be argued that pedagogical lexicography had, until the rise of the user-perspective, been a succession of more or less assumed solutions to assumed problems. This theme motivated Hatherall to undertake an observational study of dictionary-user behaviour. He criticised the popular methods of user research, such as reviews and questionnaires, on grounds of reliability problems inherent in the methods themselves. Instead, he advocated the method of direct observation of users in action, despite the problems associated with this technique.

His subjects (number not cited) were issued an introductory letter explaining the nature and the requirements of the task they had volunteered to perform. The task itself was a translation exercise of a 'difficult' text from English into German, drawn from The Sunday Times. The subjects were each given a form on which they had to indicate name, dictionary used, and what exactly it was they were looking for every time reference to the dictionary was made, regardless of whether the information retrieved was used in the translation or not. The subjects were also instructed to indicate, through codes, instances of missing words in the dictionary used, and whether they considered an item they looked up to be useful or otherwise. The task was timed for one hour, after which the subjects had to answer a short questionnaire on their strategies during the task and their attitudes and expectations.

The following results were reported (1984:187):

1. The majority of students did not read the whole text through in advance of translating, but did so after they had begun to translate and to perceive difficulties with decoding.
2. More advanced students used the dictionary more often than less advanced ones.

3. All students used only the L1-L2 (i.e., English-German) version of the dictionary when translating into German.
4. Users of bilingual desk dictionaries did not appear to look up common closed-set items such as prepositions, or open-class items which they may have met many times before.
5. All groups of students tended to translate excessively word-for-word; this was particularly true of the less advanced students.

This 'protocol' method in dictionary-user research has great potential, since it is based on the principle that data on user-behaviour is best elicited directly from the user and through direct observation of his/her performance. However, it is time-consuming and cannot be applied to a large number of users; in addition, there is no guarantee that the subjects will behave naturally under such experimental conditions.

2.18 El-Sakran (1984)

To my knowledge, this is the first study that has tackled the problem of phrasal verbs faced by ESP Arab student users of English-Arabic dictionaries. Phrasal verbs are a phenomenon in the English language that Arab learners of English find very difficult. In Arabic, the equivalents of English phrasal verbs are often single-word verbs. In addition, phrasal verbs in English consist of two or more words which may occur in different grammatical patterns (e.g., The cleaner took the tablecloth off/The cleaner took off the tablecloth/The tablecloth was taken off) (El-Sakran 1984:1).

El-Sakran chose 36 Egyptian postgraduate students pursuing various specialisations including biology (N=11), chemistry (N=6), physics (N=8), engineering (N=4) and medicine (N=7). These learners had lived in England for a time ranging from one to three and a half years. A 15-item questionnaire was given to the subjects to elicit information on dictionary ownership, frequency of dictionary use, types of information looked up, evaluation of particular dictionaries, reading of dictionary forematter, etc.

The results showed that the majority of the subjects (N=27) owned either Elias' MODERN DICTIONARY or Ba'albaki's AL-MAWRID English-Arabic dictionaries. Only two owned no bilingual dictionaries. Most of the subjects reported their dissatisfaction with Elias' dictionary for its lack of illustrative examples and insufficient information. The most popular dictionary which received high ratings, was found to be Ba'albaki's AL-MAWRID (English-Arabic dictionary). El-Sakran concluded that the bilingual dictionary used for this study was of little help to the students and argued for the inclusion of grammatical information in bilingual dictionaries for Arab learners.

El-Sakran's work was useful and original, as it attempted empirically to test the performance of a particular bilingual popular dictionary. However, the study was concerned with basically only one dictionary (i.e., AL-MAWRID) and with a small sample of foreign postgraduates studying in Britain. Therefore this may not apply to undergraduate students who are studying in their respective home countries. In addition, although the questionnaire included some useful questions such as frequency of dictionary use, the researcher, surprisingly, did not report any findings on these questions. We even were not told whether the questionnaire was administered in English or in Arabic.

2.19 Nesi (1984)

Using a lexical error-analysis approach, Nesi examined how three learners' monolingual English dictionaries - the OXFORD ADVANCED LEARNER'S DICTIONARY (OALD), the LONGMAN DICTIONARY OF CONTEMPORARY ENGLISH (LDOCE), and CHAMBERS UNIVERSAL LEARNERS' DICTIONARY (CULD), accounted for errors made by learners in speech and writing. The errors came from different sources: placement tests for postgraduate overseas students at Aston University, short oral and written tests by Algerian engineering students at Aston University and by Italian adults on holiday courses in Britain, and dialogues with French and German near-bilinguals. The errors were identified as errors of register, collocation, and meaning (Nesi 1984:18).

Pedagogical lexicographers as well as publishers of learners' dictionaries would be disturbed to know that Nesi's study revealed that the three learners' dictionaries under study failed to give the user all the information that s/he needed partly because they set out to meet an impossibly wide range of requirements. "They claim to cater for the reader, for the writer, the listener and the speaker, from any cultural background and speaking any native language" (ibid.:50).

The implication of Nesi's investigation is clear. Dictionaries need to be purpose-specific and there is a need for production-oriented dictionaries aimed at helping foreign learners of English produce appropriate and correct spoken as well as written texts. However, it may be wrong to blame dictionaries alone as responsible for language errors. The interlingual factors should be accounted for in a study like this. In addition, the learners themselves, and language teachers, may be to blame as it is possible that many learners, and to some extent their teachers, are unaware of the abundant information in dictionaries which, if properly used, might prevent the occurrence of errors.

2.20 Tono (1984)

This was an experimental investigation into the dictionary user's reference skills in which the subjects were actually asked to use dictionaries in order to observe their actual and unconscious use of information in dictionaries. Unlike Nesi, Tono's focus was on the receptive aspects of dictionary use; the subjects were asked to look up words in dictionaries and translate English texts, taken from American magazines, e.g., Reader's Digest, into Japanese. The subjects were 402 Japanese students at Tokyo Gakugei University; 63 were majoring in

English and the rest were studying Japanese education, sociology, physical education, art, elementary education, mathematics, and science. The English passages were of 100-140 words each and contained several made-up words. The reasons for including invented words were (1) to have all the students look up the same words in dictionaries, (2) to control the information in the dictionary entries, and (3) to control the lexical knowledge that the users already possessed (Tono 1984:16). The subjects were provided with specially designed dictionaries which had entries for the invented words and were asked to put the English text into Japanese by using these dictionaries.

The procedure involved testing the subjects in their own classes where they were divided into experimental groups and control groups. Some of the reported finding were:

1. The subjects tended to choose the first definition unless clear information to reject it was found.
2. The subjects tended to rely on meaning or context and used syntactic information only to reject a prima facia appropriate definition.
3. The subjects tended to look at the beginning of the entry.
4. The subjects were often discouraged to go through the whole entry because of abundant materials within the entry.
5. English majors were found to make more effective use of the information than non-English majors.

Research into strategies employed in dictionary use is lacking and this makes Tono's project particularly significant. It touches upon an area about which we know very little, i.e., the type of strategies employed by dictionary users during the look-up operation. It is also significant because, regardless of the results, it is one of the few studies that has adopted an experimental approach to the study of dictionary use. One line of research of potential interest would be into the question of whether the type of subject matter of the text and the level of text difficulty are significant factors in users' strategies.

Tono's study shed little light on the strategies or problems of the non-English majors involved. Nonetheless, the implications of Tono's work could be far-reaching from a lexicographical point of view. For example, what should be listed first in an entry? and how much information should be included in an entry? are questions which need particular attention from lexicographers.

2.21 Kipfer (1985)

A group of 292 American high-school pupils were studied by Kipfer during the year 1983-84. The aim was to examine the acquisition and influence of dictionary skills on language needs and attitudes of 'intermediate level' students. A questionnaire was used to elicit preliminary data on various aspects of dictionary use and this was followed by three tests on the reference, writing and lexical skills of the subjects. There was also an interview to complement the questionnaire.

Among the major findings were:

1. The subjects were found not to have been taught or learned dictionary skills.
2. Dictionaries were used sparingly and incorrectly and there was an unwillingness to refer to reference books for fear of appearing ignorant.
3. Dictionaries were shown to be used mainly for meaning, spelling and sometimes for pronunciation.
4. Some students were found to be unaware of the difference between major types of reference works, e.g., dictionary vs. thesaurus.
5. After subjects were familiarised with dictionaries and taught dictionary skills, there was evidence that instruction in dictionary use had positive effects on the subjects.
6. EFL learners' dictionaries were found to be of value and interest even to native speakers of English.

On the basis of her results, Kipfer argued for adopting more than one method in tackling the issue of reference skills ; she suggested three tools: (1) intensive instruction in dictionary skills, (2) promoting learners' dictionaries, and (3) exploring the potential of the computerised dictionary.

Without implying to underestimate Kipfer's thorough investigation, I should like to comment on two aspects of her methodology. First, the design of the questionnaire (1985:134) could have been better organised. There was no introduction to familiarise the subjects with the purpose, and to assure them of the confidentiality, of the questionnaire. In addition, some of the questions attempted to elicit more than one type of information at the same time; for example, question 8 reads "Have you received any instruction on how to use dictionaries? Have you taught yourself how to use them?" (1985:134). Some questions were given without specific alternatives from which the subjects could choose; for example, the questions on frequency of dictionary use did not provide the subject with a scale but were left general.

The second point relates to the implementation of the questionnaire. Kipfer (1985:11) stated that "The students were given three days to complete the questionnaires at home because the school maintained no obligation to conduct a large scale testing for an outsider.". One would wonder how reliable and trustworthy such young school pupils would be in answering the questions and what would happen when/if a particular question was not clear for some students. In my view, an 'outsider' approach raises some doubts as to the reliability of the answers, and therefore should be replaced or complemented by other 'insider' means. In my own research, a deliberate attempt was made to avoid the 'outsider' approach.

2.22 Griffin (1985)

At Southern Illinois University, Griffin conducted a pilot study among ESL students in order to develop a "dictionary skills unit". A questionnaire of 10 items was given to 128 subjects representing 13 different language backgrounds, who were divided into 4 proficiency levels from beginners to advanced according to a placement test on entrance to the EFL programme. The questions dealt with aspects of dictionary use, such as types of dictionaries owned, frequency of use, reading forematter, and other points of a general nature.

The results showed a 'lack of awareness' among the subjects as far as dictionary use was concerned, and a predominance of bilingual dictionaries as the main source of information. It was also reported that many students did not use a dictionary in class but mainly at home. There were indications that higher levels of students tended to feel embarrassed about using a bilingual dictionary frequently in class, although they might have need for it. Moreover, the majority of students did not seem to bother about the front or the back matter of dictionaries consulted.

The information gained from the questionnaire was used as a basis for designing a set of materials on dictionary-skill instruction. A group of 55 students of 9 different language backgrounds and of 4 proficiency levels was involved. The materials, based exclusively on WEBSTER'S NEW WORLD DICTIONARY, (1982 edition), consisted of a booklet containing a set of exercises based on dictionary pages. The exercises for each level were graded in terms of difficulty and transparencies were used to show dictionary conventions.

The results of the experiment showed that the subjects expressed interest and a positive attitude to the project. Of interest was:

> "A few students started with the attitude that this lesson was unnecessary because they already know how to use a dictionary, but some students, while working on the exercises, acknowledged this limited knowledge of the complications of dictionary use."
>
> (Griffin 1985:58)

I expect my own ESP students would express a similar attitude if a dictionary skills unit were to be introduced in their programme. Griffin, however, did not give any information as to the subjects' answers to question 4 on the types of information looked up most often in dictionaries used. The scope of

Griffin's study was limited, but nonetheless it added to our knowledge about the interactions between language learners and their dictionaries and demonstrated some useful design and methodology features concerning the teaching of dictionary use.

2.23 Iqbal (1987)

Iqbal has recently conducted a comprehensive study of aspects of learners' dictionaries with particular reference to the needs of Pakistani students. The aim of his study was to assess Pakistani advanced learners' reference skills and language needs. A questionnaire of 54 items covering reference skills, dictionary use, needs, and the evaluation of semantic, syntactic and phonetic information in dictionaries, was administered to 700 subjects. These subjects were all second-year undergraduates who were selected randomly from four provinces in Pakistan. The sample was assumed to be homogeneous as they were all studying English as a compulsory subject during the year 1983.

Iqbal avoided the question of ESP altogether in his thesis. He mentioned (1986:36) that "BSc students were not included on the grounds that their syllabus in English language was much more limited and orientated more towards the sciences". The findings of his survey were related to the content and format of four learners' monolingual English dictionaries, the LONGMAN DICTIONARY OF CONTEMPORARY ENGLISH (LDOCE), CHAMBERS UNIVERSAL LEARNERS' DICTIONARY (CULD), COLLINS ENGLISH LEARNER'S DICTIONARY (CELD), and the OXFORD ADVANCED LEARNER'S DICTIONARY OF CURRENT ENGLISH (OALD), to see how far these dictionaries measured up to the needs of Pakistani advanced learners of English. To complement his questionnaire, Iqbal also interviewed 25 college language teachers in Pakistan.

Iqbal's thesis is long and detailed, but the following may suffice as a summary of his major findings:

1. 92.4% of the sample possessed a monolingual dictionary of their own. However, the subjects lacked familiarity with monolingual English dictionaries.

2. OALD was found to be the most commonly used in contrast with CULD and LDOCE which were found to be used by a very low percentage (fewer than 10%).
3. Many students reported the "lack of appropriate guidance from educational institutions" as the main reason for not using learners' dictionaries.
4. The majority of students (67%) did not receive any proper advice on dictionary choice at school level.
5. Most dictionary use occurred at home.
6. Dictionaries enjoyed a very high 'image' among the student community.
7. The perusal of dictionary forematter was found to be inconsistent among the students.
8. More than half the teachers recommended learners' dictionaries to their students.
9. Both teachers and students seemed to be resigned to accepting old publications as best and were unwilling to explore new ones.
10. The teaching of reference skills had never formed part of any regular language teaching curriculum and was seldom integrated into classroom activities.
11. Meaning was the most frequently looked up information category, followed by spelling; fewer than 50% of the subjects looked up pronunciation and idioms.
12. A significant number of the students (66%) were found to be using more than one dictionary.
13. It was basically decoding activities for which the dictionaries were most often used.
14. The students were often not satisfied with the results of their enquiries in retrieving semantic information from the monolingual dictionary, because of difficulties in understanding definitions.

Iqbal suggested that user guides of dictionaries should be separately compiled for each category of L1 speaker, and recommended that instruction in dictionary skills should be integrated into all stages of the language learning process. He also proposed design features for a dictionary specifically designed for Pakistani advanced learners of English. I was impressed by the methodology and the thorough analysis throughout Iqbal's work; the size of the sample was large and the fact that the questionnaire, although rather long (19 pages), was followed up with interviews with teachers, suggested that the approach adopted was both comprehensive and valuable.

2.24 Research in Progress

In addition to the previous studies, there are currently a number of research projects in preparation. I shall briefly comment on three of the major ones.

2.24.1 Atkins et al.

Atkins et al. (1987) have reported on a research project into the use of learners' dictionaries which aims to find out "how effective a learner the student as a foreign learner is when working with a bilingual and/or monolingual dictionary" (ibid.:29). The project involves a questionnaire, in the students' own languages, and two sets of tests designed to classify them according to their English language skills and to assess their dictionary reference skills and to evaluate the content and format of the dictionaries used.

The researchers intend to include students in as many parts of the world as possible, at various levels of education. The data collected will be analysed by computer. The results of the pilot-tests were under way at the time of reporting the project. The outcome of this project should be of interest to lexicographers and to teachers who want to train their students in dictionary use. However, the scope of the project deliberately excludes ESP and restricts itself to the learning of English as a foreign language in general.

2.24.2 El-Badry

El-Badry's project concerns the use of general bilingual English-Arabic dictionaries, with the aim of designing a new dictionary. The instrument used is a 30-item questionnaire, sent to 950 potential users at universities and research centres mainly in Egypt and the U.K. A number of (almost 500) valid copies have been returned and analysed in a preliminary form. The subjects consist of 135 undergraduate students, 242 postgraduate students, and 121 people whose work is of an academic nature. Nearly half of the subjects (260) are working in the humanities field and the rest in the field of science; half of the population are resident in Arab countries, mainly Egypt, and the other half are resident in Britain.

The results so far show that the majority of the subjects (95%) indicate they need to use dictionaries, and that bilingual English-Arabic dictionaries are more frequently used than monolingual English or bilingual Arabic-English ones. AL-MAWRID (English-Arabic) by Ba'albaki has been found to be the most popular dictionary in use followed by the MODERN DICTIONARY (English-Arabic) by Elias. Meaning is reported to be the type of information most often looked up, followed by spelling and synonyms.

Because El-Badry's work is of particular relevance, most of her findings will be compared with my results throughout the following chapters. Although the two projects are different in scope, purpose and orientation, i.e., El-Badry's is concerned with general English-Arabic dictionaries for the purpose of dictionary design and does not relate to the teaching of ESP, the two projects can be considered complementary in nature.

2.24.3 Al-Besbasi

This work relates mainly to translation, but the researcher has conducted a 'protocol' experiment to investigate how translational equivalence is obtained and to explore what strategies translator trainees employ when performing the task of translation, in this case English-Arabic. The research attempted to investigate psycholinguistic aspects of the translation process using a technique which is of much potential in linguistic studies. The technique is called 'thinking aloud' and involves the subject in verbalising what s/he is doing as the task is performed. Eight subjects, both native speakers of English and of Arabic, have participated.

A short text (17 lines, 213 words) in English was given to each subject and a number of bilingual and monolingual English dictionaries were made available for use. No time restrictions were made and no restriction was imposed on the use of dictionaries. The experiment was conducted in the language laboratory and the verbalisation was audio-taped for transcription and analysis. The preliminary findings included:

1. Dissatisfaction with dictionaries was frequently expressed by all subjects; among aspects complained about were length of entries and lack of adequate equivalents.
2. Bilingual dictionaries were dominating over monolingual ones.
3. Dictionaries were used mainly for checking spelling, grammar and verifying meanings of words.
4. The subjects did not always make proper use of bilingual dictionaries as there were instances of choosing the wrong entry and missing out the appropriate equivalent.
5. There were marked differences between the subjects as to the strategies they employed. Native speakers of Arabic, for example, completely ignored using an Arabic-English dictionary although available.

CHAPTER THREE
RESEARCH DESIGN AND METHODOLOGY

3.1 Choice of General Approach

Reviewing past studies has proved to be of considerable value in considering an appropriate approach to this study. Both the design and implementation of the present project have been carried out in the light of the currently vital user-perspective approach to pedagogical lexicography and the presently popular learner-centred problem-solving approach to the teaching of ESP. The decision to employ a user-oriented approach makes sense from both the ESP and the lexicography points of view. Hartmann (forthcoming) argues "if we can develop the user perspective in lexicography as a respectable academic venture, we may see dramatic improvements even in our life-time, not only in dictionary design (in the form of special-purpose tools for both learners and translators), but also in language teaching (in the form of a greater awareness of the value of dictionary reference skills)."

Both fields share one common goal and that is the provision of help to the learner of English as a foreign language and thus the learner, in both cases, remains the receiver/consumer of an ESP input and/or a lexicographical product. However, despite the rapidly growing interest in the user-perspective, a complete profile has not yet emerged.

3.2 Choice of Type of Dictionary User

As far as ESP is concerned, the literature reviewed shows how little research has so far been carried out on the ESP learner as a dictionary user. All of the studies reviewed have been conducted outside the context of the Arab World and almost all of them have concerned themselves with non-Arab users. The Arab ESP, particularly the undergraduate, learner-user has been left untouched. Bearing this in mind, and being an ESP practitioner myself, I decided to devote this study to an ESP context within an Arab university situation.

3.3 Choice of Particular Context

The choice of the context for this research, i.e., the ESP programme at the University of Jordan, my own teaching situation, seemed to be a reasonable consideration. In Jordan, a good deal of university education, particularly in the fields of science and technology, takes place through the medium of English. The University of Jordan, the oldest and the largest in the country, has realised the important role of English ever since its foundation in 1962; all of the science faculties (Medicine, Nursing, Agriculture, Science, Dentistry, Pharmacy, Engineering and Technology) use English as a medium of instruction and in the other faculties such as Law and Economics, where Arabic is the medium of teaching, English is still deemed useful.

In 1980 the University of Jordan established the Language Centre to cater for the language needs of specialist students within the University context. Another development came in 1981 when a British Council ESP adviser was attached to the Language Centre with the function of establishing a comprehensive English teaching programme. The programme was officially completed in 1985 when the ESP adviser left and local staff took over. The first ESP client was the Faculty of Nursing which approached the Language Centre in 1981 for assistance, as its staff felt widely that the language performance of the student nurses left a great deal to be desired.

Following a preliminary needs survey, a two-year programme of ESP for Nursing was initiated. Formal and informal feedback suggested that the venture was worthwhile, and the nursing staff and students have demonstrated considerable interest and appreciation. However, the programme has been experiencing a number of challenging problems. The needs analysis was never updated, owing to time and administrative constraints, and the dictionary question was never tackled in any systematic manner.

The ESP programme does include courses for other groups of students, e.g., Law, but my choice of the nursing programme in particular was based on the following practical considerations:

1. The nursing programme is the first and the most extensive ESP activity, not only at the University of Jordan, but also, until very recently, the single largest ESP project in Jordan. Therefore, the success or failure of this programme is likely to influence the image of ESP within, as well as outside, the University of Jordan.
2. The nursing students face particular problems with English. Despite their eight years of school English, all placement tests, together with the experience of the Faculty of Nursing staff, have shown their need for tailor-made English courses. These students are not yet nurses and they are learning English and their respective subject matter, e.g., physiology, in English, simultaneously. They cannot be assumed to have developed adequate knowledge of subject specialisations upon entrance to the ESP programme and, therefore, their need for help with specialist lexis as well as general vocabulary is apparent.
3. No data are available so far about the students' reference skills and their interaction with dictionaries despite the fact that the initial needs analysis in 1981, in passing, revealed that most of the students experienced problems with dictionaries of various types.

3.4 Choice of Subjects

In the case of this study, it was decided to involve everybody concerned with the problem, i.e., the learners, the subject teachers and the language staff. Of course, there is another party involved and that is the lexicographer. In this case, the lexicographer is represented by his/her product, i.e., the various dictionaries found to be in use with the aim of finding out to what extent they meet the users' communicative needs.

The subjects involved in my research are the total population of student nurses at the University of Jordan of all levels, i.e., years 1-4, which is 500, the total population of the nursing staff (N=30), and the total population of the ESP instructors (N=8) of the Language Centre. The study aimed at covering all these three populations to serve as specialist informants as far as dictionary use was concerned.

3.5 Choice of Studied Aspects of Dictionary Use

In a recent article, Hartmann (1987) put forward four possible approaches and identified six usable techniques for research within the user-perspective approach. He did this with reference to some fifteen previous studies reported in the literature. Appropriate methodologies of dictionary use research have not yet fully been established, but research in this area can fall, according to Hartmann, into any one of the following four categories:

3.5.1 Category 1: Dictionary Typology

Under this category, a researcher may want to find out about the types of information entered in particular dictionaries. One reason that may motivate this kind of research is that language teachers are often unaware of what is in dictionaries and therefore, to their embarrassment, feel unable to recommend or choose the right dictionary for their students. Two approaches have been distinguished by Hartmann (1986a:14) as to the question of dictionary content: (1) the didactic approach which is "associated with dictionary producers and teachers and often takes the form of manuals" and the critical approach which "comes from academic and other self-appointed advisers, e.g., in the form of reviews and consumer guides".

Almost all of the reported studies on dictionary use have touched upon the issue of the dictionary typology; that by Barnhart (1962) is a major example of dictionary-typology research. In my own research, this aspect is covered (Chapter Seven) through a critical assessment of the content and format of the most popular dictionaries in current use by the students under investigation.

3.5.2 Category 2: User Typology

This category includes research into types of specific users. According to Hartmann (1981a: 302), there "will probably always be a case for the all-inclusive all-purpose dictionary, but the current trend towards specialization should not be ignored". The range of potential dictionary users is wide indeed, and this makes particular demands for a diversity of dictionary typology. Students are not teachers; student nurses are not nurses; and a foreign learner is different from a native-speaker learner. One cannot assume, therefore, that all user types share the same attitudes, preferences, interests and problems.

The foreign language learner has a special case in the market of pedagogical lexicography. Hartmann (1983b: 196) argues that "... the foreign language learner well deserves his own type of dictionary, which must be distinct from the historical, the terminological, and the translator's dictionary." Yet, there are dictionaries which still claim to cater for the learner as well as the translator and indeed every kind of user.

Much of the reported literature is based on learners of English as dictionary users; however, the ESP learner as distinct from the general English language learner needs to be distinguished as different from, though related to, the learner of general English. In this book, I shall be arguing for a new dictionary typology designed for ESP learner users as a distinct category from other types of users.

3.5.3 Category 3: Needs Typology

Establishing a needs profile of a specific group of users would help lexicographers provide more appropriate dictionary materials for the language learner. According to Hartmann (1987:28), the "range of individual needs of all... dictionary users is vast". Hartmann (forthcoming) also points out that "not much empirical research of this kind has been carried out among different types of dictionary users, so we do not yet possess a full profile of needs as conditioned by various activities".

In my study, I have attempted to incorporate a detailed needs analysis, both in general and in relation to specific contexts of dictionary use. I felt dissatisfied with the treatment of needs in previous approaches as they classified the contexts of dictionary use mainly into the traditional skills of reading, writing, speaking, listening, and translation. I have considered it more appropriate to sub-classify the needs into more specific sub-categories, such as reading textbooks, reading journals, writing technical reports, writing nursing care plans etc. This, it is hoped, will provide a clearer picture of the needs profile of this group of ESP learner users.

3.5.4 Category 4: Skills Typology

This is, according to Hartmann (forthcoming), the least researched aspect of dictionary use. Under this category, research is concerned with identifying the strategies that particular users employ during the process of referring to a dictionary. Examples of reported research in this area are Bensoussan et al. (1981), Ard (1982). Due to methodological constraints, the strategy aspect in my study was dealt with via an indirect way, i.e., the survey by questionnaire and interview, where students were asked to report on particular dictionary using habits. Criticising dictionaries alone would not solve the problems of dictionary users. This should be combined with building up students' existing reference skills in a more disciplined manner. This aspect is considered throughout this research.

3.6 Choice of Specific Techniques

Reviewing the previous literature has revealed the following investigative techniques:

Researcher/s:	Technique/s:
Barnhart (1962)	questionnaire
Quirk (1973)	questionnaire
Bujas (1975)	lexical check-list/ dictionary analysis
Tomaszczyk (1979)	questionnaire
Opitz (1979)	test/dictionary analysis
Baxter (1980)	questionnaire
Béjoint (1981)	questionnaire
Bensoussan et al. (1981)	test/questionnaire
Jain (1981)	observation/ error analysis
Ard (1982)	protocol
Hartmann (1983)	questionnaire
Mitchell (1983)	test
MacFarquhar & Richards (1983)	test/questionnaire
Chagunda (1983)	interview
Greenbaum (1984)	questionnaire
Hatherall (1984)	protocol
El-Sakran (1984)	questionnaire/test
Nesi (1984)	error analysis/ dictionary analysis
Tono (1984)	experiment

Kipfer (1985)	questionnaire/test
Griffin (1985)	questionnaire/ teaching materials
Iqbal (1987)	questionnaire/ interview/test
Atkins et al. (in progress)	questionnaire/test
El-Badry (in progress)	questionnaire
Al-Besbasi (in progress)	protocol

Most of the previous studies have used the questionnaire method; and some researchers have combined more than one method. I believe that there is a strong case for operational research techniques with particular reference to dictionary use. Most of the questionnaires used by the other researchers were presumably mailed rather than administered in the presence of the researcher. Therefore, it is difficult to take some of the reported findings seriously.

Hartmann (forthcoming) identifies six possible means through which data can be elicited regarding dictionary use, each with various advantages and disadvantages. The following is a review of them in relation to the context of my research.

3.6.1 The Critical Review

This technique has to do with dictionary typology; dictionaries are chosen for analysis of their content and format, macro and micro-structure, usually with the needs of a particular group of users in mind. The aim is usually to suggest ways of improving dictionary design and/or to help teachers and learners select a 'good' dictionary. However, this method has its problems; for example, reviewing dictionaries from a teacher's or a linguist's point of view reflects no more than his/her own experience as a dictionary user and tells us little about the real needs and skills of the users.

Nonetheless, in my research I have exploited this method by critically evaluating the suitability of the dictionaries in use in relation to the needs of the ESP learner subjects in question. The general hypothesis is that none of these dictionaries can be taken as fully reliable and adequate in meeting the specific needs of the subjects.

3.6.2 The Case Study

This is a very informative method; however, no matter how interesting the results might be, a study on a small number of learners would not yield a generalisable picture . On the other hand, we must distinguish between a small-scale case study, i.e., based on a few individual subjects and a large-scale case study, i.e., based on a large number of subjects within a specific situation. I consider my study to be an example of the latter, in that it is concerned with a population of 500 students, in addition to their subject and language teachers.

The choice of this specific group of people is intended as a case for exemplifying certain theoretical arguments and also because these students exhibit specific problems which demand specific solutions; yet it cannot be denied that the findings would be of relevance to concerned ESP practitioners and lexicographers. A study of the interaction between dictionaries and ESP learners cannot, in my view, be undertaken without taking a particular ESP programme where real learner users interact with real dictionaries in a real situation.

3.6.3 The Questionnaire

Despite all criticism, the questionnaire method will continue to be used as a major tool in any research concerned with people: what they think, what they feel, what they do, etc. It is capable of eliciting a large quantity of data about a large number of people in a short time and with less expense than other means, e.g., interviews. Above all, it lends itself easily to machine analysis, which makes the task accurate and convenient. The main problem with questionnaires is their indirectness. Nonetheless, if they are properly designed, wisely implemented, and preferably, complemented with other supporting means, e.g., interviews, they can still be very powerful in informing us about aspects of dictionary use.

In this study, questionnaires have been adopted as a major means of data collection. The rationale behind the decision to use them is that for the size of the population to be studied and for the type of data to be collected, a questionnaire would be the most suitable for the purpose. The inherent problems in this technique were kept in mind and therefore every effort was made to ensure an appropriate design as well as a suitable implementation procedure. In our ESP case in Jordan I found that we simply know very little about the students' relationship with dictionaries and therefore it was considered appropriate if we first established a data-base, i.e., a profile of various aspects of the problem upon which further studies would be pursued. Such a profile would be best obtained through a survey in which questionnaires could play a major role.

3.6.4 The Interview

With interviews it is possible to ask for explanations and to clarify obscurities for the respondent. Interviews can be open or structured, recorded or unrecorded, conducted with subjects individually or in small groups. The interview technique also has the advantage of providing an opportunity for direct observation. However, face-to-face interviews can be quite time-consuming and expensive. This technique has been employed in this research as a means to complement and verify the students' questionnaire responses and as a means to elicit the ESP staff opinions. Data obtained via open interviews can be difficult to analyse. Therefore I designed structured interviews in order to ease the analysis and achieve a match with the questionnaire results.

3.6.5 The Protocol

In this technique, an attempt is made to provide a record of the behaviour of the studied subjects. In dictionary use, this technique has considerable potential, as it is powerful in providing direct access to the strategies employed by the dictionary user. This technique, which usually uses a 'think-aloud' method and audio and/or video-taping, has not been widely exploited in the tradition of user-research. The problem with this type of technique is that it yields data only on a limited number of subjects. It also has the disadvantage of being somewhat artificial, in that the subject may not act normally under the artificial conditions imposed by this method.

In my research, I made an attempt to explore the feasibility of this method with my students. However, only indirect written 'protocol' means were used as it proved difficult, for administrative and other practical constraints, to use audio or video recording. To this end, I designed and implemented what I termed 'Dictionary Using Diaries' in order to elicit some data on dictionary use. This is illustrated in the Section 3.7.2.

3.6.6 The Experimental Test

This method has been adopted in dictionary-use research with some success (cf. Tono 1984, Kipfer 1985, Opitz 1979). In this type of method, some factors are isolated and kept under close control and a particular action is carefully measured. Experimental tests can be useful with certain types of reference tasks such as the determination of the (non-)intelligibility of dictionary definitions. No use of this technique was made in this research because no preliminary data were available. It would be inappropriate and/or difficult to conduct an experimental test without knowing beforehand for what purposes, and in which contexts, dictionaries are mostly used and which variables could influence dictionary use.

Having examined the various possible approaches and techniques, I came to the conclusion that a survey using a variety of techniques would be the most reasonable course of action.

3.7 Research Instruments: Design and Procedures

3.7.1 Pilot-Testing Procedures

Having prepared drafts of the instruments, I decided to implement them myself so that a maximum response rate could be ensured and to further pilot-test the instruments within the study environment itself. I believed that a mailed questionnaire to undergraduate students in a country like Jordan would not sufficiently yield a fully reliable picture of the studied phenomenon, i.e., dictionary use. At the beginning of the first semester in 1986/87 I started the first phase of my field-work data collection. The timing was considered to be significant as I wanted to avoid being seen as an 'outsider researcher'. Rather, an 'insider' approach was adopted to ensure familiarity with and cooperation from the subjects.

I also decided to take up two classes of nursing students to 'teach' during the whole semester in order to get an in-depth familiarity with the students' problems and to avoid being seen as a researcher rather than a teacher. The two classes were chosen randomly to be representative of the first- and second-year students attending the ESP programme. While the Dean of the Faculty of Nursing and the Director of the Language Centre were formally and informally notified about the research project's procedures and purposes, the nursing students were not told until later in the semester.

One of my research strategies was to keep a diary in which I recorded almost every event or observation, no matter how trivial it might have seemed at the time. Over the first three weeks or so I deliberately avoided giving the impression that I was conducting research where I was basically, through my class hours (3 per week for year 1, and 2 per week for year 2 students), trying to familiarise myself with the students' problems and to win their cooperation by encouraging them to come forward and discuss any language and/or study difficulties they were experiencing. I was also observing whether or not they brought dictionaries to the class and how they were coping with their lexical difficulties.

Simultaneously, I observed the ESP teachers in so far as dictionary use was concerned; for example, I checked whether or not there were dictionaries on their desks, whether there were class sets of dictionaries in their classes and whether they carried dictionaries when going to their classes. In addition, I attended almost all of the staff meetings, including those held jointly with the nursing staff, where students' problems were regularly discussed and decisions were made.

Furthermore, I examined relevant materials and design documents, syllabuses and examinations to see how the dictionary and the lexical aspects were dealt with. Meanwhile, and in the light of my observations and classroom contact with students, I was refining the questionnaires. As to the nursing staff, I conducted a number of unrecorded informal academic 'chats' in which they told me about their experience with the language problems of their students.

3.7.2 Dictionary Using Diaries

Towards the second half of the course, I distributed multiple copies of 'Dictionary Using Diaries' to my class of first-year nursing students. These were small form-like slips of paper which students were asked to keep in their dictionaries and to fill in one every time a dictionary was referred to. On each slip, the students had to indicate the date of dictionary use, the subject matter of reading they were engaged in, e.g., anatomy, physiology, the name of the dictionary consulted, the vocabulary item looked up and its meaning in Arabic.

This technique follows the 'protocol' principle discussed earlier; and the purpose of using it here was to test its applicability and feasibility and to obtain some data that may be used to complement the questionnaire method. Before implementing the 'diary' method, it was pilot-tested on ten Arab students at the University of Exeter and the University of Jordan. Originally, it was rather long and therefore time-consuming, but after the pilot-test it was made simpler and shorter so that it could be filled in within a minute or so. The students were asked to keep the multiple copies but to return the completed ones on a weekly basis. Thirteen students took part in this exercise; the findings proved to be interesting and useful, and the method itself was found to work (see Appendix 1).

3.7.3 Feedback-Based Redesign of Questionnaires

In the light of feedback, both the student and the nursing staff questionnaires were continuously modified. To enhance reliability, a draft of each of the student and staff questionnaires was typed and handed to 12 members of staff in various departments in the University, who were informed about the nature of the study and invited to provide comments and suggestions as to the format and content of the questionnaires. This 'expert panel' technique was suggested on consultation with a colleague specialising in educational research. The panel was chosen as follows: 2 from the English Department, 4 from the Language Centre English Section (two of them native speaker teachers of English of whom one was an American), 2 from the Faculty of Education whose subject specialisation was research methodology, 2 from the Faculty of Nursing, and 2 from the Arabic Section of the Language Centre. All commented on and provided suggestions for improving the various aspects of questionnaire design and its proposed method of implementation.

The same draft has also been pilot-tested on a sample of first- and second-year nursing students. My two classes together with a third second-year class were chosen for this task. 42 students simultaneously took part in this pilot-test; they were visited in their ESP classes and were told to fill in the questionnaires individually and to bring up any difficulties they experienced during the completion of the questionnaire. With another colleague I observed and noted down any instances of lack of clarity and also recorded the exact time that the students took in order to complete the questions.

It was surprising to discover that, despite all the previous efforts to ensure an appropriate design and implementation of the questionnaires, this needed yet more refinement. For example, it took the students almost 40 minutes to complete all questions, which indicated a major defect of the questionnaire, i.e., its sheer length. Various problems concerning the layout and the wording of the questionnaire were noted so that they could be avoided in the final version.

Following this in-class pilot-test and having received the comments of the panel of 12 experts, I felt a further final attempt at pilot-testing was possible. Therefore, I invited 2 groups of nursing students, each consisting of 4 students, representing the four years of study, at different times, for a small 'workshop'. The students were given, and invited to complete, modified drafts of the student questionnaire, noting any potential difficulties concerning its content and format. This workshop proved to be of considerable value as far as the questionnaire design was concerned. Some of these students, interestingly, offered suggestions, e.g., adding and/or deleting particular questions based on their own experience with dictionary use; such insights would have been impossible to obtain through mailed-questionnaires.

On the other hand, the nursing-staff questionnaire (worded in English, as it was expected to be answered by both Jordanian lecturers, who teach in English and were therefore assumed to have no problems in filling in the questionnaire, as well as some expatriate British and American lecturers) was pilot-tested on two stages. First, I discussed the content and format with one of the nursing staff and obtained some initial feedback. Next, 7 members of the nursing staff were visited as a group and asked to fill in the questionnaire. The time was recorded (a total of 15 minutes) and upon completing the task, the participants were invited to raise any problems and/or suggestions. Some inconsistencies were spotted and a few of the questions had to be completely rephrased.

3.7.4 Final Implementation of Questionnaires

Based on the feedback that resulted from the various attempts at pre-testing, a full revision of the student and nursing staff questionnaires was carried out. This was done under close and regular supervision by the Director of the Language Centre, who has long experience in applied linguistic research with reference to Jordan, and the British Teacher-Training Adviser to the University of Jordan. The final draft of the students' questionnaire (see Appendix 2) was redesigned and implemented along the following lines:

1. The language of the questionnaire was chosen to be modern standard Arabic likely to be understood by all students.
2. The length of the questionnaire was reduced to 9 pages including the introductory statement.
3. The time planned for self-administration of the questionnaire was set not to exceed 25 minutes.
4. It was decided that the implementation of the questionnaire would take place in the students' own classes with the personal presence of the researcher and colleagues from the ESP and the nursing staff.
5. The date for implementing the questionnaire was chosen towards the end of the (first) semester, but the final examinations week was to be avoided.
6. The questionnaire was aimed at <u>all</u> nursing students and therefore there was no need for any sampling procedures.

7. An opening statement introduced the questionnaire explaining its nature and purpose and inviting respondents to cooperate while assuring them of the confidentiality of their responses. (I also asked them orally not to give names or registration numbers which might indicate their identities.)
8. An official letter was sent to the Dean of the Faculty of Nursing informing her about the procedures and asking for cooperation. This was done according to the University regulations which forbid the conducting of surveys without written consent from the Dean concerned.
9. Senior nursing staff members as well as ESP teachers were deliberately involved in arranging the implementation schedule and in physically attending the sessions when the questionnaires were administered in order to give the students the impression that the survey was <u>serious</u> and that <u>everybody</u> was concerned about its outcome. This, it was hoped, increased students' interest in the study which would result in them giving responsible and, hopefully, reliable information.
10. The questionnaires were filled in by the students in class and individually. Students were invited to ask the researcher in case any point of clarification arose.
11. The questionnaire was administered on two days when the first- and the second -year students were visited in their ESP classes in the Language Centre; the third- and fourth-year students were visited in their nursing classes in the Faculty of Nursing.
12. The students were told they had the right not to take part in the survey and not to answer any question they felt sensitive about.
13. The rubrics of the questionnaire were made as clear and easy to read as possible.
14. The students were asked to <u>circle</u> round rather than <u>tick</u> in boxes because in the pilot-version it was discovered that despite a clear instruction to use ticks they

tended to use circles. When asked about this they said: "We are used to circles from our multiple-choice examinations".

15. The marginal space intended for coding, i.e., "for the researcher's use only" was omitted in the final version to save space and because the pilot-test showed that it had confused the students.
16. A five-point scale system was adopted throughout to ease coding and analysis and to save space. It had also been shown to be simple to use according to the pilot-test.
17. Special strategies were used in the questionnaire to maximise clarity and to avoid misinterpretation of questions, e.g., underlining certain words in rubrics, giving English equivalents for certain terms (for example, care plans are well-known to students more than Khitat Al-Inaya Attamridya) and examples to show what was meant by certain questions.
18. Keys (numbers and their meanings) were printed at the top of every page to avoid students' constant reference to the introduction.
19. At the end of many questions an 'any other' rubric was used to give the students a further chance to incorporate 'any other' missing item in the questionnaire.
20. At the very end of the questionnaire, as well as at the beginning, students were thanked for their cooperation.

Owing to snowy weather, only 405 out of the total 500 students of all years were present to fill in the questionnaire. On the other hand, the implementation of the nursing staff questionnaire (see Appendix 4) was carried out differently. Forty copies of the final draft were handed to a member of the nursing staff who offered to coordinate the task and follow it up. 17 copies were completed and returned within a week or so and telephone reminders were sent to the remaining respondents until 7 more copies were obtained, which left me with a total of 24 copies.

3.8 <u>The Students' Interviews</u>

The purpose of these interviews was to supplement and further enhance the reliability of the questionnaire findings. An interview as a research method has been defined as "a two-person conversation initiated by the interviewer for the specific purpose of obtaining research-relevant information, and focused by him on content specified by research objectives of systematic description, prediction, or explanation" (Cohen and Manion 1985:291). The interviews were designed on the basis of the questionnaire with more or less the same content, but with slight modifications on the format.

The advantage of the interview method over the questionnaire is that it offers an opportunity of direct interaction with the respondent and therefore allows for greater depth. In addition, the interview technique can be used to verify the questionnaire method. Furthermore, it makes it possible for the interviewer to probe the interviewee and for the interviewee to ask questions for clarification. I used the <u>structured-interview</u> type which means that the content and procedures are organised in advance and little modification is necessary because all the questions are determined in advance.

The interview (see Appendix 6) consisted of six sections: (1) background information on sex, years of study, secondary schooling and achievement in nursing and in English; (2) pre-university dictionary use on dictionary typology, contexts and purposes of dictionary use; (3) university dictionary use, including dictionary typology, sources of lexical and lexico-graphical information, contexts and purposes of dictionary use; (4) strategies related to dictionary use; (5) attitudes to dictionary use; and (6) communicative needs/problems.

A pre-test of the interview was carried out using a group of ten nursing students, and modifications were made before the final draft was prepared. In coordination with senior members of staff in the Faculty of Nursing, 41 nursing students were randomly selected using the registration lists. These students represented the four years of study as follows: 8 first-year, 11 second-year, 11 third-year, 11 fourth-year. The interviews took place over three months (each interview took around 30 minutes) and were conducted (in Arabic) individually in both the Language Centre and the Faculty of Nursing, depending on the availability of students. Because some students did not turn up for their scheduled appointments, they had to be replaced by other volunteers who were selected using the same sampling procedure. All interviewees came voluntarily and were assured about the confidentiality of their answers; no requirements were made as to the identity of these students. To simplify the analysis, the interviews were coded and the data were processed using the MINITAB statistical package. Although they proved to be time-consuming and difficult to arrange, the interviews yielded much richer and more specific information than the questionnaires.

3.8.1 The ESP Staff Interviews

The structured-interview method was also used to elicit relevant data from six (out of 8) teachers who were involved in teaching the ESP courses. The interviews, each of which took 20 minutes, were carried out (in Arabic) on a person-to-person basis and consisted of background information on nationality, experience and qualifications, the teachers' own use of dictionaries (including typology-contexts and purposes), and their attitudes as to the teaching of vocabulary and dictionary use. Despite the very small size of the sample, these interviews provided me with a further chance to acquire a clear and comprehensive picture of the ESP teachers' position concerning the issue of dictionaries in ESP.

CHAPTER FOUR
REFERENCE SKILLS AND DICTIONARY USE AT PRE-ESP LEVEL

4.1 Introduction

For university-level ESP purposes, information on the learners' previous encounters with dictionaries is certainly relevant. An ESP course is best designed when students have already acquired and developed the necessary reference skills. To establish whether or not and to what extent the students surveyed in this study had been using dictionaries, I decided to include a question in the questionnaire concerned with whether or not and how often they had used dictionaries (regardless of type or size) during the three stages of their school education, i.e., Elementary, Preparatory, and Secondary.

In the follow-up interviews, I decided to obtain more in-depth information about this aspect of dictionary use by increasing the number of questions to include the stage when dictionary use started, how it started (i.e., by what influences), types of dictionaries used, contexts in which and purposes for which dictionaries were used, and whether or not the subjects had been taught dictionary use and what they thought of such teaching. Furthermore, to achieve a clearer picture of the position of dictionaries at the pre-university level, I examined the English language syllabuses and the accompanying teaching materials for all school levels in the Jordanian school educational system. Before considering these issues, I should like to place them in a wider perspective by discussing the notion of reference skills in general.

4.2 The Notion of Reference Skills

I understand the concept of reference skills to embody the acquisition of the knowledge and ability to retrieve and use specific information from any source appropriately and with maximum efficiency. Information, verbal and non-verbal, can be sought in a variety of sources. These sources are of a wide range which includes libraries, books, encyclopedias, dictionaries, thesauruses, glossaries, computerised data-bases, telephone directories, brochures and leaflets, atlases, lexicons, and for that matter human sources of information, such as teachers, relatives and friends. Therefore, dictionary use is only a sub-category of reference skills.

Related to the concept of reference skills is the concept of study skills which is sometimes used to involve reference skills. 'Study reference skill' is a concept that, in my view, involves the knowledge and ability to extract and use specific information for learning purposes. It may be said that any attempt to obtain information is learning. This is true, but I use the term 'study reference skills' here to mean the acquisition of information in academic contexts where such skills enable the learner to learn better, i.e., cope with examinations, revise from books, take notes in lectures, and prepare seminar presentations.

The notion of study reference skills must be understood to mean more than just <u>knowledge of</u>, but also the possession of certain <u>strategies</u> that cannot be acquired by merely being lectured at but also through systematic practice and training. This is an important implication for language teaching as it is sometimes assumed that by simply <u>telling</u> learners about the content and structure of reference tools they can develop the skills to use them. It is also important to realise that by simply providing reference works, language learners, native or foreign, cannot be assumed to be able to use them properly and efficiently. This means that study reference skills need to be learnt/ taught systematically within the overall learning/ teaching curriculum; and this in turn implies that in order to teach/learn study reference skills there must be materials to teach/learn from. Such materials need to be related to specific objectives translated in behavioural terms as realisations of the attainment of the envisaged skills.

The specification of objectives entails knowledge of what is specifically involved and understood by the notion of study reference skills, which implies that such skills must be identified and broken down into sub-skills. If skills are to be taught, a methodology must also be worked out in order to achieve the set objectives. Such methodology must be principled and based on sound criteria. And finally, if objectives in a reference-skills programme are to be met, an evaluation procedure must be developed in order to assess the success or failure of such a programme.

One major goal of any educational programme is to help learners to help themselves and to train them to become independent in their learning. Building learners' reference skills is certainly an essential means of fulfilling this aim. However, this task is not always as easy as it may appear to be. For one thing, at least as far as dictionary use is concerned, while the reference skills of foreign language learners in many parts of the world are often reported to be rudimentary, dictionaries are increasingly becoming more sophisticated in content and format due to advances in technology, linguistics and language teaching (cf. Cowie 1981a:206). Therefore, language teachers should incorporate the teaching of reference skills in their designs, methods, materials and tests, and pedagogical lexicographers should orientate their dictionaries to the levels of proficiency and the specific needs of target learner groups.

Target dictionary user groups can be identified according to their lexical needs and their reference skills, i.e., retrieving and using lexical information appropriately and efficiently (Cowie quoted in Béjoint 1981:207). However, the specification of reference skills of a particular group of target learners is a complex task. This is because the available knowledge on the look-up reference strategies is still limited due to lack of research in this area (cf. Hartmann 1987:23). Yet, regardless of the type of dictionary used, the following points about the nature of dictionary reference skills can be made.

Firstly, the reference activities of a dictionary user consist of two stages which match the two structures of the dictionary: the macro-structure and the micro-structure. This implies the need to teach the following strategies:

(a) finding the desired lexeme, by locating the right headword or the right subdivision of the entry;

(b) extracting the specific piece of information needed (e.g. meaning, spelling, pronunciation (cf. Béjoint 1981:211)

Secondly, the look-up operation is not a simple mechanical activity. Scholfield (1982a:185) argues that successful dictionary use involves a number of steps that should be performed correctly and requires from the learner-user prior knowledge of certain language rules and dictionary conventions. In the process of looking-up, the learner-user is engaged in testing hypotheses and making inferences in relation to the item in question and its context. Therefore, dictionary use is an interactive process which requires a set of strategies. Scholfield (ibid.) postulates that the dictionary look-up process in the context of reading comprehension involves seven steps:

1. Locating the unfamiliar item/s.
2. In case of inflected items, removing the inflections to recover the form to look up.
3. Searching for the unknown item in the alphabetical list.
4. In case of not finding one main entry,
   (a) trying to look up each main element if the unknown item is a fixed phrase, idiom, or compound;
   (b) trying the entry for the stem, if the unknown item has a suffix;
   (c) scanning nearby entries, if the unknown item is an irregularly inflected form;
   (d) searching the addendum if there is one.
5. Reducing multiple senses or homographic entries by elimination.
6. Understanding the definition and integrating it into the context in which the unknown item is met.

7. If none of the senses entered fits, trying to infer one that does from the available senses; if more than one sense fits, searching for additional clues from the context in the source text.

Thirdly, the reference skills in dictionary look-up operations vary according to the direction involved in language use, i.e., decoding (listening and reading) and encoding (writing and speaking). According to Hartmann (forthcoming), dictionary use in reading involves a 'semasiological' process which involves the user's ability to recognise and locate words and understand their meanings in context. In writing, on the other hand, an 'onomasiological' process is involved where the writer needs to find a word or phrase appropriate to the concept to be verbalised. In the act of translating, both processes are employed.

Fourthly, the dictionary reference skill consists of the following basic components (Hartmann forthcoming). These are:

(a) Recognition of a need, i.e., acknowledging a lexical problem.
(b) Decision to consult a dictionary, which assumes familiarity with reference materials.
(c) Locating the headword, which assumes familiarity with the alphabet or the dictionary macro-structure.
(d) Selecting the desired item from the dictionary entry, which involves knowledge of the dictionary micro-structure and the ability to choose the most relevant sense.
(e) Integration of the chosen item into the original context.

Finally, dictionary reference skills are bound to vary according to the user's age, attitudes, memory, linguistic proficiency, level of study, familiarity with a particular dictionary, and the availability of particular dictionaries (Hartmann ibid.).

4.3 The Place of Dictionary Use in the School English Curricula

For the purposes of this work, the focus will be mainly on dictionary aspects rather than on other components such as structure or phonology. Both Curricula are published in English and the quotations used here are verbatim.

4.3.1 The Compulsory Stage

The English Curriculum for the Compulsory Stage (Grades 5th and 6th Elementary and 1st, 2nd and 3rd Preparatory) has been adopted by the Ministry of Education since 1965/66. The section on 'aims' and 'specific objectives' makes no mention of the acquisition of reference skills and/or the ability to use reference tools of any sort. Therefore, it can be argued that this Curriculum does not recognise the achievement of the ability of using reference materials and particularly dictionary use amongst the stated objectives in teaching English as a foreign language. However, in advocating the 'Oral Direct Method', this Curriculum provides some suggestions for the teaching of vocabulary.

Apart from the general notes and clues as to methodology of teaching vocabulary, no indication is given of any role that dictionaries or other reference works might play in the teaching of vocabulary. The only direct reference to dictionaries and reference materials occurs in the section 'Co-curricular Activities':

> "6. Encouraging library visits for the purpose of looking up words and information in dictionaries, encyclopaedias, atlases and other reference works."

The implication of this is quite clear. Dictionaries and other reference tools simply have no place in the curriculum. Pupils and teachers can only be encouraged to visit a library whenever they want to look up something. After all, the very word 'co-curricular' essentially implies that such activities are only of peripheral importance to the curriculum proper.

I also searched through the course textbooks used in the Compulsory Stage to find out whether there was any input on dictionary use and/or reference skills in general. The only input was found in New Living English for Jordan: Book Three (taught to First Preparatory Class). The 'Tenth Lesson', which is the first lesson to be taken at the beginning of the second semester, starts with a list of vocabulary items to be presented through a short reading passage (± 150 words) which under the title of 'A Dictionary' defines the word 'dictionary' and gives some information on dictionary typology, then it introduces dictionary use. To practise the alphabet a rhyme is given. This is followed by 'Dictionary Work' which is a list of English words, with their meanings in English, together with illustrative drawings. The learner is required to "Read them, learn them, and do the three exercises on them". Comprehension questions on the reading passage follow the list and the three dictionary-related exercises mentioned above. The lesson ends with a dictation exercise. Thus, throughout the five years of school study in which English is taught for six periods per week, pupils receive just one lesson on dictionaries and their use, and only one aspect is practised, namely the alphabet. The rhyme exercise is interesting and certainly a useful method for practising the alphabet provided that the teacher makes sure that his/her pupils understand what the bold-face words stand for. The comprehension questions based on the reading passage are mechanical and would not engage the learner in an enjoyable activity. Little quizzes such as the one incorporated in the dictation exercise are potentially valuable in so far as they arouse the learners' curiosity to use a dictionary. However, the last sentence in the dictation passage reflects the authors', and, to a great extent, the syllabus designers' low regard for the dictionary, and leaves the learner with the impression that the only place for dictionaries is in libraries.

Having said all this, it is fair to mention here that over the past few years the scene of English language teaching at the Compulsory Stage has witnessed dramatic and significant changes. In 1984, the Ministry of Education published the new Compulsory

Stage English Curriculum, designed by a specially commissioned team of applied linguists and educationists from the Ministry of Education, the University of Jordan, and UNRWA (United Nations Relief and Works Agency). This document came about as an attempt to update and improve the quality of English language teaching in Jordan in view of the recent developments in psycholinguistics and applied linguistics. The rationale given for launching this new syllabus is seen as a response to "national and international developments in education" (p.6).

Having analysed this document, it is possible to make the following observations:

1. It recognises the role of dictionary use as an integral aspect of the syllabus, unlike its predecessor which considered use of dictionaries as merely a 'co-curricular' activity.
2. It clearly regards dictionary use as part of the 'specific objectives' within the teaching of reading.
3. It explicitly advises the teacher to train pupils in the use of a learner's monolingual English dictionary as part of the methodology of vocabulary teaching.
4. It realises the importance of vocabulary teaching and allocates it a separate section, while reference is made to the teaching of vocabulary in all other sections.
5. It specifically suggests the use of a learner's monolingual English dictionary and makes no reference to the use of bilingual dictionaries.

However, this curriculum gives no indication as to how dictionary use may be taught; it also provides no suggestions regarding the evaluation of the achievement of the objective relating to dictionary use. In addition, the authors do not explain why the dictionary is only introduced at the First Preparatory Class. Moreover, the attitudes of the authors towards the use of bilingual dictionaries are not clear although they suggest the use of an English-English dictionary.

It must be added that the newly introduced curriculum has led to the production of a set of interesting and more useful materials. In this connection, I examined the two course books which were implemented on a trial basis in 1987. These books are:

1. PETRA : Progress in English Through Relevant Activities for Jordan; Students' Book 3. Trial Edition, 1987.
2. PETRA : Progress in English Through Relevant Activities for Jordan; Workbook I. Trial Edition, 1987.

My purpose in examining these books was to determine how the dictionary input was treated in these materials and to see how similar or different the new materials were to the old ones. It must be recognised that both in content and in format the activities offered in the materials accompanying the new syllabus for the Compulsory Stage are a major step forward as far as instruction in dictionary use within the school level is concerned. To say the least, the dictionary practice exercises, among other aspects, are nicely presented and carefully integrated into the syllabus. It will be interesting to see how these new materials work as the 'trial' editions of these books become established. In conclusion, it must be realised that the recent integration of the dictionary in the classroom is, if properly handled by teachers, a very positive move in Jordanian English language teaching.

4.3.2 The Secondary Stage

In 1971 the first curriculum for the teaching of English at the Secondary level was published by the Ministry of Education. Before that time there was no syllabus to be followed and the only way to get any idea on the English curriculum was to infer it from the textbooks and examinations used. In order to establish whether or not provision was made for dictionary practice input at the Secondary Stage, I scanned the 50-page English Curriculum : Secondary Stage, first edition 1971. This Curriculum was produced in anticipation of the currently used textbooks and at the time was describing plans for the future rather than realities. The 'Introduction' (pp.4-5) mentions five requirements for the preparation of a 'teachers' handbook' which was supposed to accompany the textbooks. The second of these requirements reads as follows:

> "2. The handbooks should make recommendations on the proper use of the English-English dictionary. At appropriate places in the textbooks, dictionary exercises should be included."

This indicates that the Curriculum designers recognised the role of the monolingual dictionary in the learning process. Yet, no mention was made of the learner's ability to use a dictionary of any type within the 'General Aims' or the 'Specific Objectives' (p.9).

The fifth 'specific objective' (p.10) is interesting, in particular. It mentions the desirability to "acquire the linguistic skills and techniques needed for advanced work at post-secondary levels". Although vague in nature, this objective implies that, at the Secondary Stage, pupils should be trained in those skills and techniques that will help them cope with the requirements of work at college and university levels. One wonders whether the acquisition of reference skills, including the use of dictionaries, is not worthy of mention within these 'specific' objectives.

Unlike the new curriculum for the Compulsory Stage (1984), the Secondary Stage English curriculum makes no mention of vocabulary teaching and dictionary use under its headings and sub-headings. There is a section on 'Co-curricular Activities' (p.26) where, surprisingly, even the single item about "encouraging library visits for the purpose of looking up words and information in dictionaries, encyclopaedias, atlases and other reference works" is missing. Nowhere else in this curriculum does instruction in dictionary use enjoy a mention. Finally, I searched through the course books assigned to the Secondary Stage and failed to trace any 'dictionary exercises' referred to in the introduction of the curriculum.

From the above remarks, it may be concluded that dictionary use and instruction in reference skills receive no better attention at the Secondary Stage than at the Compulsory Stage. The curriculum at the Secondary Stage is still structure-based and waiting to be functionalised, and in need of a radical reform (along principles and features) similar to the one introduced in the new curriculum for the Compulsory Stage. Throughout the three years of English teaching at the Secondary Stage (5 periods a week plus 2 additional periods for the Literary Stream pupils only, p.3), pupils cannot be assumed to have received any formal tuition in dictionary use and reference skills. One can only hope that, in succession to the new syllabuses and materials recommended for the Compulsory Stage, the present syllabus and materials for the Secondary Stage will be replaced by new ones as a natural and logical continuation of the Compulsory Stage.

4.4 Findings on the Students' Use of Dictionaries at School Level

In the students' questionnaire I asked the students how often they used dictionaries (regardless of type or size) during the three stages of their school education, i.e., the Elementary, Preparatory,and Secondary. The frequency of dictionary use was indicated by reference to a 5-point scale ranging from 'never' to 'always'. Throughout this book only three frequency indicators will be used as follows: 'low' (rarely), 'average' (sometimes) and 'high' (often) and (always).

Table 1: Dictionary use at school.

Stage of Schooling		Frequency low	average	high
Elementary	N=	77	73	48
N=198, %=48.8	%=	(38.8)	(36.8)	(24.2)
Preparatory	N=	62	152	148
N=362, %=89.3	%=	(17.1)	(41.9)	(40.8)
Secondary	N=	58	110	200
N=368, %=90.8	%=	(15.7)	(29.9)	(54.3)
(Each N in this column is out of 405)				

As Table 1 shows, less than half the student sample reported that they used dictionaries at all at the Elementary Stage. More than a third of them who said they did use dictionaries indicated a 'low' (rarely) frequency of consulting dictionaries. The picture changes dramatically if we look at the Preparatory Stage where it seems that the majority of students (89.3%) reported using dictionaries and under half said they used a dictionary with a 'high' frequency. At the Secondary Stage, both the numbers of students who reported using dictionaries and the number of 'high' frequency users seem to increase sharply.

In the follow-up interviews, more information was collected on this aspect of dictionary use. In reply to the question whether they had been using dictionaries before joining the University, 39 students out of 41 said 'yes', and only two replied 'no'. Thus, most of the students in the sample seem to have been dictionary users throughout their school attendance. Asked at what Stage and/or Class they had started using a dictionary, out of those 39 students nearly half (48.7%, N=19) said they began using a dictionary at the Preparatory Stage; 12 students (30.7%) indicated the Elementary Stage as the beginning of dictionary use, whereas at the Secondary Stage only 8 students (20.5%) said they did not remember. As for the Class at which students started using dictionaries, 13 out of 37 students reported the First Preparatory Class (age 12), and 4 said they did not remember.

Table 2: Stage of starting dictionary use.

Stage	No. (out of 39)	%
Elementary	12	30.7
Preparatory	19	48.7
Secondary	8	20.5

This can be explained by reference to the old and new Curricula of English which, as we have seen earlier, introduce the dictionary at this particular stage.

Students were also asked how they started using dictionaries and under what influences/factors. The results were as follows:

Table 3: How dictionary use started at school.

Factor	Yes		No		No answer	
	N	%	N	%	N	%
1. Teacher of English asked the class to use a dictionary	19	46.3	17	41.4	5	12.1
2. Teacher of English advised the class to use a dictionary	20	48.7	15	36.5	6	14.6
3. Others (parents, friends, etc.) advised me to use a dictionary	18	43.9	15	36.5	8	19.5
4. Personal initiative (I decided to use a dictionary)	17	41.4	12	29.2	12	29.2

(Numbers are out of 41)

These figures show that more than one factor influenced students' decisions; Almost half the sample said that the teacher of English did either ask or advise them to use a dictionary, or that family members and classmates encouraged them to do so.

Information was also obtained as to the types and sizes of dictionaries used before joining the University. Here is a summary of responses:

Table 4: Types and sizes of dictionaries used at school.

Dictionary Type	Dictionary Size	Used by	
		N (out of 41)	%
English-Arabic	pocket	29	70.7
English-English	pocket	8	19.5
Arabic-English	pocket	1	2.4
English-Arabic	large	23	56.0
English-English	large	17	41.4
Arabic-English	large	-	-

This table clearly indicates the students' reliance on English-Arabic dictionaries. It is also of interest to report that bilingual Arabic-English dictionaries were reported to be almost out of use; a finding which calls for an investigation into the reasons for this phenomenon. A tentative explanation might be the lack of adequate dictionaries of this type and also the fact that dictionaries at school level are mainly needed for decoding purposes.

Furthermore, the interviewed students were asked to report names of particular dictionaries they used during their schooling. Table 5 shows which dictionaries were said to have been used and by how many.

Table 5: Dictionaries in use at school.

Name of Dictionary	Used by N (out of 41)	%
AL-MAWRID (English-Arabic)	23	56.1
Elias' MODERN DICTIONARY (English-Arabic	24	58.5)
DICTIONARY OF SCIENTIFIC AND ENGINEERING TERMS (English-Arabic)	1	2.4
OXFORD ADVANCED LEARNER'S DICTIONARY (English-English)	15	36.5
COLLINS ENGLISH DICTIONARY (English-English)	1	2.4
WEBSTER'S NEW COLLEGIATE DICTIONARY (English-English)	1	2.4

The above table confirms that students rely on bilingual English-Arabic dictionaries during their school education. It is interesting, however, that more than a third of the students said they used OALD. What explains this finding is that the English examination in the G.S.E.C.E. (General Secondary Education Certificate Examination) usually includes some questions that demand reference to a learner's monolingual English dictionary. This does not imply that dictionaries are allowed in the examination or that there are questions on dictionary use, but rather that the English examination has items on spelling, stress, suffixation, and word-pronunciation. The bilingual dictionary does not usually cover much of this information, and therefore students preparing for this examination commonly refer to OALD which is recommended by teachers of English. This observation was confirmed by the students interviewed.

The main activity in which students used dictionaries at the pre-university level was reported to be reading comprehension; 95.1% (N=39) of the sample said they used their dictionaries for understanding school textbooks in English. Using dictionaries for composition came next, as reported by 56% (N=23). Only 17% (N=7) said they used dictionaries for listening purposes (e.g. watching television programmes in English), and only one student (2%) reported having used a dictionary in order to speak. As to the type/s of information looked up in dictionaries, the majority (95.1%, N=39) reported meaning. Spelling came second, reported to have been looked up by 17 (41%.4) students. 12 students (29.2%) said they looked up pronunciation, and 9 (21.9%) claimed to look up grammatical information. Eight students (19.5%) said they also sought general-interest information in dictionaries.

I specifically asked the students whether or not and at what Stage and/or Class they had been taught how to use dictionaries during their English lessons throughout their school learning. Less than a third of the sample (31.7%, N=13) said they had received guidance in dictionary use; 69% (N=9) of these 13 students indicated that this instruction had taken place during the Preparatory Stage, particularly in the First Preparatory Class. The students were also asked whether they remembered what types/names of dictionaries they had been taught to use. The results showed that these were mainly English-Arabic pocket dictionaries; only 2 students said they had been shown how to use English-English dictionaries. The students also reported that the type of training they had received was basically one aspect of dictionary use, i.e., finding headwords using the alphabetical sequence. In addition, eleven students (84%) out of 13 said the instruction received was 'not enough'; although 9 of them (69%) reported that they thought the instruction they had received was 'useful'.

In concluding this chapter, I should like to reiterate that the acquisition of reference skills and systematic instruction in dictionary use at school level is essentially relevant to the teaching/learning of ESP at the university level. Students at the university stage are supposed to be independent in their learning and to have a choice in making their own decisions. Using dictionaries properly can lessen their dependence on teachers, remembering that neither the ESP teachers nor the subject teachers have enough time to teach everything. Having examined the situation of dictionary reference skills at the pre-ESP level. I can conclude that the nursing students cannot be assumed to possess any reliable reference skills of any sort.

Although there have recently been signs of improvement with regard to the place of dictionaries in English curriculum, the situation seems to be still rather unsatisfactory. At the Secondary Stage there is literally nothing worthwhile mentioned on dictionaries and at the Compulsory Stages (Elementary and Preparatory) the new syllabuses and materials are still being trialled. Furthermore, it is not enough to incorporate dictionary practice exercises in the textbooks and to mention dictionary use specifically as an objective. It is certainly a positive move, but it should be coupled with teacher awareness of dictionaries. This implies that teacher-training programmes should make sure that the dictionary input is handled by teachers who know dictionaries and have positive and objective attitudes towards them.

Teachers of English should also be aware that merely telling students, in passing, about the dictionary is not the whole story about dictionary skills. Language teaching planners should also realise that the place of dictionaries is not only in the library, but also in the classroom. Learners should have access to class sets of learner's dictionaries and should be trained in how to use them, in class, and through systematic exercises and interesting activities.

Moreover, the findings showed that despite the recommendation in the English Curriculum that training should be given in using a monolingual English dictionary, students still seem to rely too heavily on English-Arabic dictionaries. It is generally accepted that bilingual dictionaries are not necessarily 'bad' and that learners cannot easily be 'weaned away' from them. Yet, learners' monolingual English dictionaries can be more useful provided that users are made to realise this and given adequate training.

It is a well known fact that learners in Jordan place high value on examinations. It is worthwhile considering the possibility of including dictionary use (knowledge and skills) in examination questions to make sure that both teachers as well as students take this into account. Most important of all, I should like to propose that the Ministry of Education English language planners envisage a nationwide research project into the position of dictionaries and reference skills at the school level. Research in this area is seriously lacking and the advantages of such research are predictable. Among the potential results of this type of research could be a learner's dictionary for Jordanian school pupils.

CHAPTER FIVE
CONTEXTS OF DICTIONARY USE IN RELATION TO COMMUNICATIVE NEEDS

5.1 Introduction

Nowadays, few experts would dispute that both the teaching of ESP and the making of dictionaries should be needs-sensitive. Whether our purpose is to develop ESP learners' reference skills or to design appropriate ESP reference tools, an up-to-date awareness of the specific communicative needs in the target learning environment is a pre-requisite. In this chapter, the contexts in which dictionaries are said to be used by the student nurses at the University of Jordan are related to a detailed profile of these students' communicative needs.

A full section in each of the students' and staff questionnaires and the follow-up students' and staff interviews was devoted to collecting data on communicative needs and problems. In addition, some of the subjects' attitudes towards various aspects of dictionary use and the students' strategies in relation to dictionary use are incorporated throughout this chapter. To place this in a wider context, I shall first discuss the tertiary educational system in Jordan and the role that English plays in it. Particular attention is given to the status of English in the University of Jordan and to the place of dictionary skills in its EFL/ESP programmes.

5.2 The Role of English in the Jordanian Tertiary Education

Perhaps due to its lack of material resources, Jordan found the natural alternative in a considerable investment in developing its human resources. Higher education is clearly a key factor in this development. Recently the Ministry of Higher Education was created as the body responsible for planning and running further and higher education in Jordan. This has come in response to the rapidly increasing demands by Jordanian people for higher education. In 1987 the number of Jordanian students studying abroad was approximately 40,000 distributed over 50 Arab and foreign countries, (Al-Shabab magazine, 1988, 146:11); also the increasing number of Secondary-school graduates has led to an increase in the number of higher education institutions. For example, while the number of community colleges and polytechnics has risen to over 50, the number of universities has increased to four, and in addition a private national university is planned. These universities are: the University of Jordan (in the centre of the country), the University of Yarmouk and the University of Science and Technology (in the north), and the University of Mu'tah (in the south).

Few people in Jordan would doubt the need to learn English in higher education. This is not surprising, as English has established itself as a major international language of science and communication. In business, in tourism, in industry, and in learning/teaching at higher education institutions, a good command of English has indeed become a must, as well as a prerequisite for securing jobs. With the recent increase of unemployment in Jordan, even among doctors and engineers, proficiency in English is clearly an advantage in the job market. In fact, few of the many newspaper advertisements for vacant posts do not specify English as a requirement. People in Jordan want to learn and use English not only as a means of getting jobs, but also of doing their jobs more efficiently. For practising professionals, e.g., doctors, nurses, engineers, lawyers, businessmen, a good working knowledge of English means keeping up to date with what is going on in the constantly changing modern world.

At college and university levels, the need for English is more than a bare necessity, as the language is used as a medium of instruction. Such a need is quite clear among students in science and technology departments where most of the communication, spoken and written, is in English. There is, for example, the need to understand spoken English in lectures, seminars and in practical demonstrations delivered by Jordanian as well as overseas staff. In addition, students are often referred to textbooks, handouts, manuals and journals written in English; and in many cases, students are also expected to write their examination answers, technical reports and research essays in English.

Compared to the school level, the position of English teaching in further education is relatively better off. English is usually taught at this level by better qualified staff, often with a master's degree in EFL and/or applied linguistics, plus several years of teaching experience. In community colleges and polytechnics, where Arabic is the main medium of instruction, English is taught as a compulsory subject, though with varying degrees of duration, frequency and weight. Having taught for some time in community colleges, I believe that the standards of English teaching at this level are still unsatisfactory, especially as far as ESP is concerned. The concept of ESP is still new in the country, and despite the fact that there are courses taught under the name of ESP, Jordan still lacks expertise in this area. However, there are signs of positive change. At the moment, discussions are going on, involving the Ministry of Education and the Ministry of Higher Education, with the aim of improving the current syllabuses, materials and methodology.

At the university level, English currently enjoys a strong position despite the increasing campaigns for arabicising higher education. This is because English plays a special role in the learning/teaching process. In other words, English is essential in this context for students and staff, particularly in science and technology departments, for academic purposes. In all four Jordanian universities, science and technology subjects are still conveyed through the medium of English, and in the human/social sciences departments English is considered necessary wherever there are courses that must be taken in English. Moreover, each of the four universities has its own Department of English where B.A. and M.A. programmes in English language and literature are offered, and/or its own Language

Centre where service ESP courses are provided. The staff teaching English at university level usually have Ph.D and/or M.A. qualifications in various areas of linguistics and EFL plus relevant teaching experience.

5.3 The Teaching of English at the University of Jordan

The University of Jordan, and particularly its Language Centre and Faculty of Nursing, are the specific contexts within which this research has been conducted. The University of Jordan was established in 1962, almost sixteen years after Jordan's independence, as the first national institution of higher education. Now, it comprises on its campus 14 faculties, 10 specialised centres and a well-stocked library; more than 12,000 undergraduate and postgraduate students pursue their studies in a wide range of subjects.

The University policy-makers have long realised the importance of and the need for English. The following observations reflect the University of Jordan's policy towards English:

1. The medium of instruction in all science faculties (i.e., Medicine, Nursing, Dentistry, Pharmacy, Agriculture, Science, Engineering and Technology) is English.
2. English is regarded as a necessary requirement in the faculties of human and social sciences (i.e., Law, Arts, Economics and Administrative Sciences, Physical Education, Education, and Shari'ah or Islamic Studies); even though Arabic is the medium of instruction in these faculties, there is hardly a study plan that does not incorporate one or two courses to be taught in English.
3. Within the Faculty of Arts, there are undergraduate and postgraduate programmes leading to the degrees of B.A. and M.A. in English Language and Literature.
4. The Language Centre was established in 1980/81 with the main objective of providing a comprehensive service-

English programme to fulfil the practical needs for English as a means of achieving academic and professional purposes.

5. One year after the establishment of the Language Centre, a four-year ESP project was initiated and a British-Council ESP adviser was attached to the University with the aim of building up and developing the project.
6. In 1984, and as a follow-up to the ESP programme, a new four-year EFL teacher-training project was set up within the Language Centre. A British teacher-training specialist helped establish this programme, which is now a comprehensive postgraduate Diploma/Master's course in TEFL run in co-ordination with the Faculty of Education and the Department of English in the University. The candidates for this programme are usually in-service teachers and supervisors of English working in both the public and the private sectors.

The English Section of the Language Centre aims to fulfil both a service function, i.e., identifying the needs and solving the problems of English use within the University environment, and a training function realised through the various programmes and courses directed at specific groups of learners. These programmes include:

1. The University Programme, which is aimed at the wider student population; consisting of compulsory and elective courses. The compulsory credits include (a) English 100, (b) English 99 and (c) the UPE (University Proficiency Examination). It is a regulation in the University that every student must pass the UPE, and take either English 100 or English 99, according to performance in the UPE, in case s/he fails the UPE. The elective programme is a set of undergraduate and postgraduate courses based on specific needs of students in various departments.
2. The ESP Programme, which is based on the particular needs of individual faculties and departments in the University. This programme includes a two-year

programme for nursing students, and a two-course programme for law students.

3. The Teacher-Training Programme, which involves postgraduate Diploma and M.A. courses designed with the purpose of training in-service Jordanian secondary school/and community college teachers and supervisors of English.

5.4 The Place of Dictionary and Reference Skills in the University EFL/ESP Programmes

With the purpose of finding out whether or not, and if so, to what extent dictionary and reference skills have been incorporated in the University EFL/ESP programmes, I examined a number of course files in the Language Centre. These files included course design documents, teaching and test materials, and administrative minutes. I also sought out the lexicographical input in a number of course outlines and course descriptions available in the Faculty of Education and the Department of English.

At the undergraduate level I discovered that none of the University compulsory courses, i.e., English 100 and English 99, had a lexicographical component. Ability to use dictionaries was missing in the list of objectives and there was no sign of any dictionary tasks in the teaching and evaluation materials being used. Neither did the UPE involve questions on dictionary skills. What was interesting was that the University Proficiency in Arabic (an equivalent university graduation requirement targeted at the Arabic Language) did include a section on the ability to use monolingual Arabic dictionaries. What was also interesting was that, from personal communication with colleagues in the Arabic Section of the Language Centre, I learnt that there was usually a high failure rate on the dictionary component of this examination, which suggests that university students do not acquire the necessary skills of using mother-tongue dictionaries. The UPE in English and the courses related to it represent, from the University administration point of view, the minimum proficiency in English that any university student cannot do without. This means that in the present system many students graduate without any evidence to show that they are capable of handling reference materials in English.

On the basis of research carried out by the English Section of the Language Centre during 1982-85, a University-wide elective programme was proposed and approved. Due to shortage of staff and other administrative problems this programme has not yet been implemented. The programme consists of six courses: Proficiency in Spoken English, Using Academic Journals, Principles of Report Writing 1, Principles of Report Writing 2, Advanced English Vocabulary, and Study Skills in English. The last two do mention reference skills. For example, one of the aims of the Advanced English Vocabulary course is "helping students to understand and use general principles of lexical selection in English ... with work focussing on, for example, ... selection of appropriate items or meanings from reference materials (thesaurus, monolingual dictionary etc.)". The Study Skills in English course also mentions "use of reference materials" among its objectives. This is certainly a sign of a positive change in the direction of integrating reference skills into university courses; yet the courses in this programme are optional; unless students and teachers in the University are made aware of their value the effectiveness of the programme may be very limited. In addition, there are still no teaching materials, design documents and suggested methodologies as the programme is very new.

As far as the ESP Programme is concerned, I scanned the course files relating to the English for Students of Nursing and English for Students of Law programmes. The ESP for nursing students is a two-year programme that is required by the Faculty of Nursing for all its students and is studied concurrently with the students' medical and nursing subjects during their first two undergraduate years. It consists of an entrance placement test and four one-semester courses of study. The component courses include: Foundation Course: Talking about Nursing, Managing Lectures, Handling Textbooks, and Reports and Seminars. Although I found a statement in the Preparatory Design Document mentioning that the nursing students are neither proficient in using bi- or monolingual dictionaries, I failed to find any dictionary input in any aspect of the programme, i.e., course descriptions, objectives, methodology, teaching materials (published or locally produced), tests and administrative minutes.

The results of the students' questionnaire confirmed the above-mentioned finding as more than two thirds of the sample (69.1%) disagreed with the statement: "The present English language courses provide sufficient information on dictionaries."; only 13.8% agreed whereas 17% could not make up their minds.

The Language Centre also offers two undergraduate ESP courses on behalf of the Faculty of Law. These are: Legal Expression in English, and Techniques of Legal Drafting in English; none of the course descriptions and the related materials makes any reference to the ability of using reference materials.

At the postgraduate level, the Language Centre runs a Diploma/M.A. programme in TEFL in cooperation with the Faculty of Education and the Department of English. The courses that are the responsibility of the Language Centre are: Language Skills 1, Language Skills 2, Methods of Teaching EFL 1, Methods of Teaching EFL 2, Language Variation, Language Description, and Teaching of English for Specific Purposes. The courses offered by the Department of English and the Faculty of Education, respectively, are non-EFL, i.e., theoretical linguistics and educational subjects. None of the above-mentioned courses makes any mention of lexicography and/or the teaching of dictionary use and reference skills either in the syllabuses or in teaching materials.

It may be relevant here to mention that throughout my four-month counterpart training (October 1987-January 1988) within the University of Jordan Teacher-Training Programme, I came to notice that the lexicographical awareness of the majority of the postgraduate in-service teacher-trainees left much to be desired. For example, during one of my team-teaching lessons I asked the Diploma/M.A. TEFL class whether they had to refer to a thesaurus in preparing one of their assignments. Everyone said they did not know what a thesaurus was and when I showed them a copy of ROGET'S THESAURUS no one recognised it. Having experienced this I contacted the University main bookshop, the principal supplier of books written in English in the country, and I was asked what I meant by the word thesaurus!

Outside the Language Centre, English is taught mainly in the Department of English and in the Faculty of Education (as a sub-specialisation). I looked at the Study Plan and course descriptions of the Department of English (from which I graduated) to seek out the lexicographical components. The vast majority of courses seem either to be biased towards theoretical linguistic subjects or literature-oriented. Lexicography has no status as a subject in the curriculum and EFL teaching is given only one course, which is optional. Most of the graduates of the Department of English end up teaching EFL at school and college levels in Jordan and therefore it can be assumed that the majority of the present English language teachers in Jordan have received little tuition in lexicography and reference skills.

As far as the Faculty of Education is concerned, I looked at two handouts relevant to the sub-specialisation in the teaching of English. The handouts provide information on the nature, objectives, methodology and syllabus of two courses, i.e., Advanced Reading, and Vocabulary. There is no mention of dictionary use and reference skills in the former, but the latter does specify using monolingual English dictionaries 'with skill and ease' as its last objective in the list of aims. Students on this course are also required to compile a 'word-card dictionary' of the new vocabulary they learn with the meanings and usages of each word (however, this task counts only for 5% of the final grade), and are recommended to purchase a 'good' English-English dictionary.

From this review of the position of dictionaries and reference skills in the University of Jordan, it is possible to conclude that apart from the course on vocabulary within the Faculty of Education sub-specialisation in English, and the study skill component in the proposed elective programme, the lexicographical input does not feature in any of the University EFL/ESP courses at any level. It also seems that deciding to incorporate a lexicographical element in the University courses is left to individual teachers rather than part of a set policy. The situation is clearly crying out for reform; and it is one of the aims of this research to make a contribution to this.

5.5 An Analysis of Communicative Needs

5.5.1 The Notion of Needs Analysis

It is axiomatic that establishing a profile of the ESP learner's communicative needs is an essential point of departure in any ESP activity. However, to conduct a needs analysis is by no means a simple task. For one thing, the notion of learner needs itself is susceptible to varying interpretations. For some, 'needs' may be seen in terms of passing particular classroom examinations; and for others the concept of needs may encompass real situations where language is of vital importance for survival. According to James quoted in Chambers 1980:25), the concept of needs analysis involves "real and imaginary needs, demands, requirements, expectations, and use". There is at the moment a lack of precision in ESP as to what the term 'needs analysis' exactly involves.

Munby's work (1978) represents one of the most rigorous attempts in ESP to provide a working procedure for specifying the needs of particular ESP learner groups for the purpose of syllabus design. He developed a 'processor' of communicative needs and provided an 'instrument' which could help an ESP practitioner to build up a profile of his/her students' needs.

But the Munby model has been received with much criticism by ESP academics. For example, Widdowson (1981:2) has argued that it is wrong to assume that a specification of language needs should define the language content of a course. The concept of 'learner needs' is, according to Widdowson, open to at least two interpretations: it can refer, on the one hand to "what the learner needs to do with the language once he or she has learned it" which is a 'goal-oriented' understanding of needs. On the other hand, the term may imply "what the learner needs to do to actually acquire the language", which is a 'process-oriented' definition of needs. On this basis, Widdowson (ibid.:9) suggests that we should consider the parameters in the learning process and the specific methodologies of science rather than just the specification of linguistic goals (cf. Hutchinson and Waters 1987).

Unless we are concerned with a homogeneous group of learners, the needs-analysis task can prove to be very difficult because different groups may not have identical needs. Another problem in needs-analysis procedures is that the learning needs are not always stable and can change in a short time due to administrative and socio- and psycho-linguistic factors. Information on needs should not be elicited from learners only, but rather from a combination of sources. In the case of this research a variety of techniques were used to obtain relevant information about the students' needs from several sources, i.e., questionnaires, interviews, classroom observations, tests, documents and written 'protocols'. These techniques were used to obtain information from the learners themselves, the subject staff, and the ESP teachers involved.

The term 'needs analysis' means not only collation of information, but also categorising and establishing priorities. It can thus have two meanings: (1) the analysis of needs or (2) analysing in order to establish needs (Chambers 1980). In my study, I employ the latter definition of the term. In other words, I do not take the term 'needs analysis' to be concerned only with linguistic specifications but also with other aspects of the target situation, including the students' profile, the learning environment and the overall academic context in which the students are learning ESP (cf. Swales 1980a).

5.5.2 The Subjects of the Study: A Profile

Since the 1981 initial small-scale needs analysis, the profile of the ESP programme for nursing students has never been updated. The following paragraphs provide an update:

5.5.2.1 The Nursing Staff

In the session 1985/86 the number of staff members was 31, 18 of whom were teaching assistants (B.Sc. holders), 12 assistant lecturers (M.Sc. holders) and one assistant professor (Ph.D. holder); there were also 14 members of staff on study-leaves. Recently two members of staff of American and British nationality joined the Faculty. The nursing staff questionnaire was completed by 24 members of staff and yielded the following background information on the staff:

(a) 22 (91.6%) were Jordanians and one was not (one did not indicate nationality).

(b) The staff's teaching experience at university level ranged from 1 to 5 years, 8 (33.3%) said they had one year of experience, 5 (20.8%) two years, 5 (20.8%) three years, 4 (16.6%) four years, and 2 (8.3%) had five years. Therefore, the majority of staff members seem to be relatively young and of limited experience.

(c) The previous finding was confirmed by the data on staff qualifications which revealed that the vast majority, 19 (79.1%), said they had a B.Sc. degree, 3 (12.5%) an M.Sc. and only one reported having a Ph.D. qualification.

(d) Most of the staff members (75.0%, N=18) said they had studied at the University of Jordan for their most recent academic qualification and only 4 (16.6%)

reported receiving their most recent qualification overseas.

(e) The following courses were reported as being taught: Pediatrics, Medical Surgical Nursing, Maternity, Community Nursing, Health Education, Nursing Administration, Nursing in Our Lives, Fundamentals of Nursing, Obstetrics and Gynaecological Nursing, Child Development and Care, Trends in Nursing, Research Methods, History of Nursing, Anatomy, Physiology, Pharmacology, Biochemistry, Microbiology, Physics, Chemistry, Sociology, and Psychology.

(f) All members of staff said they were teaching more than one level of students, i.e., first year, second year etc.

(g) As far as their own use of English in their teaching was concerned, all the teachers reported using English, but in varying degrees: 13 (54.1%) reported that they used English with high frequency (often), 9 (37.5%) said that they used it 'sometimes' and only 2 (8.3%) reported a low frequency of using English, i.e., 'rarely'. This suggests that although English is the medium of instruction in the Faculty of Nursing, some of the teachers rely on both English and Arabic in their teaching.

(h) Related to the previous finding is the nursing staff's attitude towards the use of Arabic in teaching nursing subjects. The results showed that the vast majority (87.5%, N=21) said they would disagree with the statement "Nursing subjects should be taught all in Arabic"

This shows that the majority of the nursing staff value the role of English in teaching nursing topics highly, despite the fact they have to resort to Arabic occasionally, depending on the difficulty of the topic and the level of students' understanding. Concerning their own use of dictionaries, all of the 24 nursing staff said they used dictionaries in preparing their lessons; 9 (37.5%) with 'often' frequency, 10 (41.6%) with 'sometimes' frequency, and only 5 (20.8%) said they used dictionaries 'rarely'.

5.5.2.2 The ESP Staff

Out of the 6 ESP teachers interviewed, 5 were of Jordanian nationality. Apart from the one non-Jordanian Arab teacher, all of the ESP staff reported having taught at university level for a period of no more than 6 years. The academic qualifications of the ESP staff were at the level of B.A. in English Language and Literature from Jordanian universities (except the non-Jordanian teacher), and at the level of M.A. in TEFL/TESOL, and English Literature from British and American universities; and only one had an M.Sc. in Teaching ESP from the U.K. All of these teachers were involved in teaching more than one of the ESP nursing courses.

In general, the ESP staff suffer from status problems, i.e., they are not regarded by the University administration as academic staff but rather as 'employees' or 'instructors', which means that they do not usually enjoy the same prestige and job benefits as their colleagues in other departments. This creates problems of motivation, which in turn affects the standards of teaching at all levels. Another problem is that few of the ESP teachers seem to have received ESP-oriented training; something that results in problems in running the various aspects of the ESP programme.

I asked the ESP staff about their own use of dictionaries. All said they consulted dictionaries fairly frequently when preparing their lessons, mainly for meaning, pronunciation and spelling of technical terms. 4 of them reported checking usage in their dictionaries, but only 1 said that he made use of etymological information in dictionaries. All said they used their dictionaries in reading whereas 4 said they used dictionaries in writing activities. 3 ESP staff reported using dictionaries in a listening comprehension situation, mainly watching television, and none reported using dictionaries while engaged in a speaking situation. As to whether or not they carried dictionaries to their classes, only 2 said they occasionally did so, and one of the reasons mentioned was: 'I am not used to this'.

I also asked the ESP teachers whether or not they had received formal training in dictionary use, general reference skills, or lexicography during their undergraduate and postgraduate studies. Only 2 said they had had some lexicographical input during their training; they indicated that though such input was useful, they felt it was not adequate. It may be interesting to note that when asked which dictionaries they thought their students were using, 3 teachers said they had no idea; the other 3 could only guess that the AL-MAWRID (English-Arabic) DICTIONARY and HITTI'S MEDICAL (English-Arabic) DICTIONARY were in use.

5.5.2.3 The Nursing Students

The nursing students represent a homogeneous group of ESP learners; they all share similar features in that they are of the same nationality (almost all are Jordanians), of the same age group (18-22 years), of the same cultural background (Jordanian Arabs), with the same mother tongue (Arabic), of similar educational backgrounds (graduates of government schools), in the same social and physical setting (University of Jordan), and studying for the same educational purposes (B.Sc. in Nursing).

The majority of the nursing students are female. Out of the 405 students who answered the questionnaire, female students constituted 67.6% (N=274). The follow-up interviews confirmed this finding as out of the 41 students in the sample, there were 26 female students (63.4%). It is a policy of the University and the Faculty of Nursing to admit a majority of female students. Until 1982/83 the Faculty accepted only female candidates but, understandably, due to the fact that Jordan still suffers from a shortage of qualified nurses and because of the relatively low social status of the nursing profession in general, the University has started to encourage male school graduates to join the profession.

According to the Faculty official records, the number of all undergraduate students during the year 1986/87 (first semester) was 510, divided as follows: 150 first year, 120 second year, 106 third year, and 134 fourth year. In this study, 405 students of all undergraduate levels participated. The high response rate can be held to be due to the following:

1. The questionnaires were collected immediately after the students had completed them;
2. some nursing and ESP staff members were present helping with the administration of the questionnaire;
3. Both nursing students and staff showed a high level of interest in the study.

The four levels of study were represented in the study as follows (one student did not mention his/her level and therefore the total is 404):

Table 6: Levels of representation in the sample

	N	% out of sample population (out of 405)	Level of Participation %
First Year	127	31.3	84.6 (out of 150)
Second Year	96	23.7	80.0 (out of 120)
Third Year	79	19.5	74.5 (out of 106)
Fourth Year	102	25.1	76.1 (out of 134)
Total	404		

These numbers show that each of the four levels of study was adequately represented; the relatively lower level of participation of the third-year students could be explained by the fact that it was a rainy and cold December day when a group of the third-year students were scheduled to complete the questionnaire. In the interviews (41 students), each level was represented by 11 students except the first-year level which was represented by 8 students only, due to timetabling problems.

Authenticity of teaching materials in ESP is a debatable issue, and the question: what is meant by authentic (the text, the task etc.)? is still controversial. However, ESP practitioners have to decide what type of materials their students prefer to learn from. From my own ESP experience, I came to realise that a balance and a variety of materials would be the best approach, as it was found that many students complained when they were taught from purely medical/nursing materials equally as when they were taught from non-medical/ nursing materials. To have an idea of what students and nursing staff think about this issue is, I believe, important if dictionary-based reference materials were to be designed for the nursing students. The overall picture indicates that the majority of students appreciate the notion of relevance in ESP materials. Responding to the statement "It is better to learn English through nursing-related texts/ subjects", 67.6% of the students answering the questionnaire said they agreed, while 25.1% of them disagreed and 71% were undecided. More than two thirds of the nursing teachers (N=17, %=70.8) agreed with the similar statement: "English should be taught to nursing students through nursing-related materials.", whereas 7 (29.1%) disagreed.

One characteristic of the nursing students is that they tend to value examinations highly and therefore expect a direct link between the content of teaching materials and the content of final examinations. Because these students are examination-oriented, they tend to attach great importance to assigned textbooks as a main source of input from which they revise and prepare for examinations. A learning mode commonly adopted by the students is rote learning and memorisation.

It is also the Language Centre's experience that the students are very interested in linguistic information in their English courses; although they tend to enjoy <u>doing</u> things while learning, they insist, at the same time, on a body of knowledge as an accumulation of facts rather than the acquisition of skills. The information-oriented attitude among the students is presumably due to the nature of the academic environment in which they are learning their subject courses. Students are expected to demonstrate their knowledge of the discipline in terms of passing information rather than skill-based examinations (cf. Crocker 1982:8). As far as preferred modes of learning are concerned, the nursing students tend to enjoy working in pairs or in small groups; in addition, they tend to prefer variety in the types of activities from which they are to learn ESP.

In general, the majority of the nursing students can be considered as 'lower intermediate' as far as their overall level of proficiency in English is concerned. By this I mean that the majority of these students on entry:

1. can write only single sentences at most. They generally lack the rhetorical skills of text organisation and adequate control over the conventions of writing, e.g., the use of punctuation. In one hour they can be expected to write approximately one paragraph of simple sentences of approximately 100-150 words on a familiar, general topic. They can write simply on a limited number of everyday situations, but make many errors in spelling and structure. In general, they lack control over lexical choice and their writing is stylistically simple.
2. They can form single utterances about their main intentions, but are usually inaccurate when expressing their opinions, and generally lack the linguistic knowledge necessary for communicating precise ideas. In general, they cannot maintain a conversation for more than a few minutes. They make frequent pronunciation (stress and intonation) errors, but can be generally understood by a native speaker because of contextual clues and paralinguistic means. Their speech is slow and characterised by repetitions and hesitations.

3. They can handle short (±150 words) written texts of a general nature, but face difficulties in identifying meanings of unfamiliar technical and general lexis. They are able to read factual texts with the help of a bilingual dictionary, but cannot make inferences, although they generally comprehend the overall gist of the text. Their reading is quite slow and they are usually inefficient in using reading strategies, e.g., skimming, scanning, locating specific details.
4. They are generally weaker in listening comprehension than in reading comprehension, presumably because at school the teaching and testing of listening skills is neglected. They can follow general-topic conversations but not without difficulty. In general, they lack specific listening skills such as note-taking and using contextual clues. They find it quite difficult to comprehend spoken English on subject-specific topics delivered by native or near-native speakers. They may be able to identify the overall idea in a discourse, but they usually fail to retain specific details and tend to ask for repetition and clarification. (I based my definitions of the students' level of proficiency upon personal experience and upon D. Wilkins' 'Proposal for Level Definitions' in J. Trim "Some possible lines of development of an overall structure for a European unit/credit scheme for foreign language learning by adults", Council of Europe Documents, Strasburg, 1978, pp.71-78.)

5.5.3 <u>Resources and Constraints</u>

An essential component of any ESP needs-analysis activity should be devoted to resources and constraints within the learning environment. Time seems to be a rare commodity in the ESP market where everyone concerned, i.e., ESP staff, subject staff, and learners, is bargaining for more time for his/her needs. In the Faculty of Nursing both teachers and students are under considerable time pressure as their timetable is usually full of academic, administrative, and practical clinical training duties. Originally, the ESP programme was intended to cover the four years of study and the time agreed upon was three hours per week for every course. Now the number of courses and the amount of time conceded for ESP have been reduced to cover only the first two years of study and, apart from the first course, the frequency of contact hours is only two hours per week. Throughout the structured interviews and in the students' responses to the open-ended item on the questionnaire, the fact that the time for the ESP courses was not enough was emphasised. Some students even mentioned that they did not use dictionaries because of time constraints.

In total, throughout the ESP courses, the students receive no more than 95 teaching hours (3 hours x 15 weeks + 2 hours x 15 weeks x 3 semesters) if we exclude examination periods. This is by no standards enough time to build the students' proficiency up to a reasonable level, remembering that the majority are 'lower intermediate' when they join the ESP programme. In this respect, serious thought should be given to the possibility of developing self-study materials to supplement the contact hours. If study/reference skills were to be introduced into the ESP programme, it would seem unreasonable to argue for an additional separate independent course, as it would be unlikely that the Faculty of Nursing would accept any more pressure on their timetable.

The availability of classrooms is a problem often mentioned. Both the Language Centre and the Faculty of Nursing lack sufficient and appropriate classrooms; in many cases the ESP students have to be accommodated in large auditoria, never intended for language teaching purposes. Not only is the availability of classrooms a problem, but also their physical design (e.g. fixed small seats), which does not allow for interactive group/pair work.

There is only one audio-visual room in the Language Centre and even this is often shared with other departments and usually booked as a classroom for other non-ESP courses. There are four language laboratories but these, too, are shared and often used as classrooms; in addition, many of the machines are usually out of order. There is only one video set (no video cameras are available) and it will be a long time before computers can be introduced into the University language teaching format.

ESP textbooks are generally available, but as no one textbook is perfectly appropriate they are often supplemented by locally produced materials; however, there are restrictions on the photocopying and duplication of materials. The Language Centre does contain a good cassette collection of EFL/ESP materials but it was my observation that much of the material remains unknown to the ESP teachers and consequently the ESP students (for example, I accidentally discovered that the collection contains oral supplementary materials on how to use the LONGMAN DICTIONARY OF CONTEMPORARY ENGLISH; not a single person seems to have used it).

The administrative status of the ESP nursing courses seems to be the main problem at present. Many students stated explicitly (through interviews and the open-ended item in the questionnaire) that the pass/fail uncredited nature of the courses was behind the high rate of absenteeism and the low level of students' motivation in studying ESP. A similar problem relates to the training of the ESP teachers themselves; most of them had not received any ESP training and their own career status seems to be shaky in comparison with that of their colleagues in other departments. Some students appeared to be very critical about the ESP staff: they complained about the content and methodology adopted and even offered suggestions for improvement (among such suggestions was the possibility of introducing an 'all-difficult-words-booklet' for each level of students together with audio-taped pronunciation exercises for these 'difficult words').

As far as the availability of lexicographical materials is concerned, there are no class sets of dictionaries available to the ESP students; it is apparent that dictionary acquisition is left to the learners themselves. I scanned the reference collection of the Language Centre and found that while a good number of EFL books was stocked, the lexicographical component was very small. However, most ESP teachers seem to keep their own copies of dictionaries on their desks.

The Main Library in the University of Jordan is well stocked with monolingual and bilingual dictionaries and other reference materials, e.g., encyclopedias and atlases. However, it is doubtful whether the ESP students, and probably teachers, are fully equipped with adequate library skills. In addition, the time pressure and the fact that the Faculty of Nursing is physically located a considerable distance from the Library make it difficult for the students to make proper use of the Library's facilities. Both the students' questionnaire and the follow-up interviews referred to this problem. Almost half of the student sample (47.4%) said they never used these facilities. In addition, 40.8% of the students who said they did, reported doing so with 'rare' frequency.

The results of the student interviews confirmed the above-mentioned finding. Out of the 41 students in the sample, only 7 (17.0%) said they used Library dictionaries. However, double this number, 14 (34.1%), said they used the dictionaries available in the Faculty of Nursing library. Indeed, a number of the nursing students interviewed were totally unaware that there were dictionaries available in the University Library at all. This leads me to raise the whole issue of library use and the interaction between students and libraries in general, a topic worthy of a separate investigation, in my view. At present, there are no courses offered on library skills. Therefore, in the design of ESP courses there is a need for a component relating to library use.

The University of Jordan Bookshop is the main supplier of books for students and staff. Having browsed through its shelves, I found that both bilingual English-Arabic and mono-lingual English dictionaries were stocked, but there were no thesauruses available. The same observation was made upon visiting a major book exhibition held in the University of Jordan and organised by a group of bookshops in Jordan.

5.5.4 Dictionary Use and Linguistic Needs

Language is a tool for communication and learning. Communication and learning can be facilitated by the appropriate use of dictionaries; therefore dictionary use is a means rather than an end in itself. Our interest in dictionaries in ESP is justified insofar as dictionaries can help ESP learners meet their language needs and solve their communication problems. Communication is a two-way process, i.e., encoding and decoding a written or a spoken message which involves a variety of linguistic activities performed in a number of contexts.

For the ESP practitioner and the pedagogical lexicographer, it is of value to know in which communicative situations the ESP learner is engaged while consulting a dictionary. Are dictionaries being used primarily in receptive activities such as reading and listening or in productive tasks like writing and speaking? In the literature on dictionary use, the use of dictionaries is often related to the four language skills (i.e., reading, writing, speaking, and listening) and translation. In my view, it is insufficient to speak of dictionary use in terms of general language activities as it is possible that even within the same basic language skill such as reading, the user may be using dictionaries in varying degrees for different purposes, e.g., reading textbooks, journals.

The students were presented with ten possible communicative contexts and were asked to indicate whether and to what extent they used dictionaries (regardless of type or size). One blank line was provided at the end of the list and the students were requested to add any other context not mentioned in the questionnaire. As for the nursing staff, the same contexts were presented in the questionnaire and subjects were asked to indicate the importance they attached to the students' use of dictionaries in each context. Opinions of students and staff as to the need for help in using English in each of these contexts were elicited through the questionnaires and the interviews.

The list of communicative contexts was arrived at after direct consultation with a number of nursing staff and students of various levels before the questionnaires and interviews were finally administered. In addition, to make sure that the students understood what was meant by each item on these lists, English equivalents and examples were provided in writing, and linguistic and TEFL jargon was avoided in the wording of the questions. The following is a taxonomy of language needs related to dictionary use. It must be noted here, however, that the subjects' responses incorporated here may reflect their perceived attitudes rather than their actual practices; nevertheless, the data reported can serve as useful indicators of general trends.

5.5.4.1 Decoding Needs and Dictionary Use

Decoding involves language use in reading and listening comprehension activities. Understanding the meaning of unfamiliar words is one of the most important aspects of comprehension.

5.5.4.1.1 Reading Comprehension

To understand medical and nursing written materials in English can be a complex task. It is therefore essential to build up the readers' ability to understand inter- and intra-sentential relationships in a text, and to make use of contextual clues in order to work out unfamiliar vocabulary. An often emphasised reading strategy in the EFL/ESP literature is contextual guessing as a means of identifying vocabulary meaning and of minimising the need to use 'time-consuming' learning aids such a dictionaries (cf. Kruse 1979, Phillips 1983b, Vanparreren and Schouten-Vanparreren 1982, Bramki and Williams 1984, Clarke and Nation 1980, Bensoussan and Laufer 1984). Both the students' questionnaire and interviews contained a question on the extent to which the students claimed to have made use of the strategy of guessing unfamiliar word meanings from context while reading and before deciding to consult a dictionary. The vast majority of students (95.5%) reported using the strategy of guessing, about two thirds of them (66.6%) reporting a 'high' frequency of doing so. Out of the 41 students interviewed 36 (87.8%) said they used the strategy of guessing before using dictionaries. These findings suggest that the vast majority of the sample attempt some sort of guessing while reading and that despite the overall consistency among the four levels of students, there seem to be some variations in the frequency of using this strategy (e.g. the second-year students appear to guess less frequently than the first-year ones). It would need to be shown, however, how successful such students are at guessing and how guessable the vocabulary items they usually encounter are.

The implication of these findings for the ESP teachers is that they should capitalise on the students' willingness to use this strategy, incorporate it as an explicit objective in the design of the ESP course and train the students how to be successful guessers. The notion of guessing assumes that there are clues in the context that can help the reader to interpret the unfamiliar items in a text. It is not clear though whether certain words are more guessable than others. Bensoussan and Laufer (1984) conducted an interesting piece of research on lexical guessing in relation to reading comprehension in English. The study revealed that "context helped lexical guessing in only 13 percent of the responses for only 24 percent of the words" and that "word guessability was shown to be less a function of using the context than of applying 'perceived notions'". The study also showed that "although more proficient students know more words than less proficient students, they were not able to use context more effectively" (ibid.:15).

The implication of Bensoussan and Laufer's study is that guessing alone is not sufficient because of the possibility that reading texts may embody unknown words which are essential for comprehension and which cannot be worked out from the context, possibly for lack of sufficient clues. The other implication is that unless students are properly trained in the skill of guessing and made aware of its limitations, there is a danger that these students misinterpret the text due to 'perceived notions'., e.g., knowing one meaning of a word and sticking to it even when it is wrong because of the polysemous nature of the word.

I do not wish to underestimate the role of teaching guessing strategies as a pre-dictionary activity, but I think that with particular texts and certain levels of learners a reasonable use of dictionaries can be a safer strategy than mere guessing. Nonetheless, before the learner decides to use a dictionary, s/he should be taught to exploit and maximise the use of the following contextual aids:

1. word components, e.g., prefixes, suffixes and stems
2. pictures, diagrams and charts
3. parentheses and footnotes
4. synonyms and antonyms
5. exemplification clues
6. summarising clues
7. the syntactic functions of the word (i.e. parts of speech)
8. orthographic conventions (punctuation and capitalisation)
9. collocations of the word
10. cohesive devices and logical connectors
11. lexical familiarisation (intentional non-dictionary type of definitions in a text cf. Bramki and Williams 1984).

I have found some evidence that, although the majority of the students in the sample reported attempting guessing before dictionary use in reading, with a 'high' frequency, they tend to look up all unfamiliar items. 345 students (85.1%) reported looking up all unfamiliar items when reading; 44.9% of those students said they did so with a 'high' frequency.

During the data-gathering period, I noticed that some students tended to gloss (in Arabic) the meanings of items they looked-up, in the texts they read. To explore the students' use of this strategy, specific questions were entered in both the questionnaire and the interviews. The results showed that an overwhelming majority of students (92.8%, N=376) reported using this strategy; more than two thirds of them said they did so quite often. The interviews also showed that out of 41 students, 35 (85.3%) said they used this strategy. The striking observation is that this strategy seems to be most popular amongst the first-year students which suggests that they are facing more lexical problems than the other groups.

For the ESP practitioner as well as the ESP lexicographer, the implication of this finding is that getting access to a representative corpus of texts used by students would provide opportunities to study the typology of lexis that the students usually look up, which is itself an indication of vocabulary problems and needs. The finding that the first-year students appear to use this strategy more frequently than any of the other three levels indicates a relationship between study level and the lexical needs on the other hand.

Having observed that some students were glossing the meaning of items they looked up, in notebooks especially for the purpose of building up their own 'dictionaries' of newly acquired lexis, I was interested to find out to what extent this strategy was popular among the students' sample. Unlike the strategy of glossing in text, this strategy seems to be generally unpopular (except amongst the first-year students) as less than half the sample (45.1%) said they used to gloss in special vocabulary note books, and less than a third of such students (31.1%) reported a 'high' frequency.

In the questionnaire, the students were asked whether or not and, if so, how much need they felt for help in practising reading comprehension using nursing/medical texts. On the whole, the need for practising reading comprehension via subject matter texts seems to be evident amongst all levels of students and particularly among the first-year group. The vast majority of the students (89.8%, N=364) said they needed help in this respect, out of these students almost half (48.6%) wanted 'much' help whereas 30.2% said their need level was 'average'.

5.5.4.1.1.1 Textbooks

Here I need to make a distinction between the need to understand textbooks which are usually purchased by students at the request of subject teachers and the need to understand library books which are usually suggested for further reading. It is the former type that needs to concern us here as it is this type of textbook which students value highly since such textbooks serve as the major source of information after lectures and because their content usually features in examinations.

The overwhelming majority of students (91.3%, N=370) across all levels said they needed help in understanding textbooks. However, the degrees of need indicated show that the fourth-year students felt that they needed less help than other levels of students.

In addition, 17 members of the nursing staff out of 23 (73.9%) reported a high level of need for help, and 6 (26.0%) reported an 'average' degree of need for help amongst the students as far as understanding textbook English was concerned.

As for the use of dictionaries in understanding textbooks, the results showed that the vast majority of the students (93.5%, N=379) said they were using dictionaries in the context of reading textbooks; the majority of them (61.4%, N=233 out of 379) reported a 'high' frequency of dictionary use in this respect. On the other hand, the interviews showed that 40 students out of 41 said they were frequent users of dictionaries in the context of reading textbook materials.

To assess the role of dictionary use in reading textbooks, the students were requested to indicate their opinions as to the statement: "Nursing students can generally understand the content of their assigned textbooks without reference to a dictionary". A majority of the students (65.1%, N=264) disagreed with the above statement, which suggests that students to a great extent rely on dictionaries in their attempts to understand their textbooks. As for the nursing staff, the findings showed that 13 (54.1%) out of 24 said they thought it was 'very important', and 7 (29.1%) said it was 'extremely important' for their students to use dictionaries in reading textbooks.

Textbooks constitute a major component in the academic life of students. In fact, the nursing students at the University of Jordan are required to buy certain textbooks which are in English and usually designed for either American or British native speakers of English. However, I was told on a number of occasions by students as well as nursing staff that the students are not expected to read and understand the whole of a textbook. Instead, they are usually assigned specific sections/chapters to read; these are used as follow-up study activities to lectures and/or seminars and as revising materials for examination purposes. Textbooks are also used as source material when students are asked to prepare oral presentations and technical reports.

The previous findings are worth considering by both the ESP course designer and the ESP dictionary maker. The language of nursing and medical textbooks has received little attention from discourse analysts. Therefore, I should like to emphasise that a representative sample of nursing/medical textbooks in use by Arab students should form an important component in any database from which ESP dictionaries and/or ESP materials will be drawn in future.

5.5.4.1.1.2 Handouts

Lecturers usually depend on handouts for conveying information to their audiences. This area of medical/nursing English seems to have been missed out in the literature on discourse analysis; yet, the majority of students (90.8%, N=368) across all levels reported the need for help in this area although the amount of help required seems to be less than that required in the case of textbooks (only 24.7% wanted 'much' help whereas 43.4% needed 'average' help). Out of 23 nursing lecturers 16 (69.5%) reported a 'high' degree of need to help students understand handouts.

The fact that the nursing students face lexical problems in handouts can be seen by reference to the considerable number of words looked up, and glossed in Arabic, on the handouts students use as a source of information. The interviews also revealed that 56% (N=23) of the students reported using dictionaries in looking up items in handouts regularly. As to the nursing staff, out of 24 teachers, 14 (58.3%) said that it was important for their students to use dictionaries in order to understand handouts.

5.5.4.1.1.3 Specialist Journals

ESP students need to understand the content of specialist journals in English although this need is not as immediate as that of understanding textbooks, for example.

Research on linguistic features of medical journal articles has been reported by Baker (1988); and Adams-Smith (1983) who, interestingly, has pointed out that medical journals demonstrate the deliberate use of jargon words and pile-ups of complex nouns by medical writers, which can be very difficult for non-native readers.

A large percentage of the students (89.8%, N=364) said they needed help in this area; at least one third of them (34%) described their need as 'much'. Among 23 nursing staff, 18 (78.2%) said the students' need for help in understanding journals was 'high' and 2 (8.6%) said the need was 'average'.

As far as dictionary use is concerned, 259 (63.9%) students said they used a dictionary while reading journals but the frequency of use (20.8% reported a 'high' frequency) seems to be much lower than in the case of textbooks, for example. The percentage of dictionary users in this context seems to be increasing as the students move up the academic scale. Understandably, the highest numbers of users are noticed among the third- and the fourth-year students as it is these students who are more likely to be required to refer to specialist journals.

Out of 24 nursing staff members, 19 (79.1%) said it was 'very important' for their students to use dictionaries in the context of reading journals. This suggests that reading specialist journals is an area where students face considerable lexical problems. For the ESP teacher as well as the ESP dictionary designer a sample of journal articles in medicine and nursing should be considered for inclusion in the linguistic written data upon which specialist syllabuses and dictionaries will be based.

5.5.4.1.1.4 Written Examinations

To understand the English used in written examination papers is of particular interest to students. This is because examinations play a significant role in the educational system in Jordan. The use of dictionaries in examination settings is a debatable issue. For some EFL/ESP teachers dictionaries should not be allowed in examinations while for others they are seen as harmless if not useful. Research by Bensoussan et al. (1981) showed that there was no significant correlation between students' use of dictionaries and their results on a reading comprehension test. However, Bensoussan (1983:342) believes it is possible to integrate dictionary use in EFL reading tests and that it is possible to construct questions which take into account that students will be allowed to use dictionaries during the test. Motivated by Bensoussan's work, I explored the subjects' attitudes to the use of dictionaries in both ESP and subject examinations. No less than 50% of the students in general are against the idea of allowing dictionaries in ESP tests. On the other hand, the interviews showed that 19 students (48.7%) out of 39 said it was 'very necessary' to have dictionaries allowed in ESP tests, while 6 students (15.3%) thought this was 'necessary'; and 11 (28.2%) were against the idea. The interviews with the ESP teachers also showed a wide discrepancy of attitudes to this issue. Out of the 6 teachers, 3 were opposed to allowing dictionaries in ESP tests on the grounds that the dictionary is time-consuming and that the dictionary may provide answers to some questions. The other three agreed, but with strong reservations; for example, they said dictionary use should be considered, taking into account the type and content of the test and the presumed level of students' proficiency in the language.

The students' views on the use of dictionaries in subject tests indicate a similar lack of agreement. More than half of each of the first- and the second-year students agreed with the statement "Nursing students should be allowed to use dictionaries during nursing examinations" while more than 50% of the third-year and the fourth-year students disagreed. The results of the students' interviews confirmed that at least 51.2% (N=20 out of 39) were in favour of allowing dictionaries in examinations in nursing subjects. In contrast, while 9 out of the 24 nursing teachers (37.5%) agreed that nursing students should be allowed to use dictionaries in nursing tests, 62.5% (N=15) were against the idea.

For the ESP teachers as well as the nursing teachers the implication is that dictionary use during examinations should be made optional as almost 50% of the students prefer to be allowed to use dictionaries in examination contexts. Dictionaries provide information about individual words but such information on its own cannot provide full comprehension of texts or writing an argument. The information provided by the dictionary can only help in the process of arriving at particular solutions to particular problems. Unless the questions in an examination are ultimately asking for meanings of words, the dictionary can be of no harm, if not useful. For the ESP lexicographer the implication is that a list of words commonly used in academic tests such as discuss, comment upon, analyse, compare, explain, should be entered in a dictionary designed for academic purposes, and clear guidance should be provided on the meanings and use of such words. This task can be facilitated with the help of specialist informants.

It may be relevant to compare the findings of this study with the findings of some other researchers concerning the use of dictionaries in reading comprehension. For example, El-Badry (in progress) found that the vast majority of her subjects 93.2% said they used their dictionaries in reading activities, and for more than a half of them (57.5%) dictionary use in reading was said to take more than 50% of the total dictionary consultation. Iqbal (1987:82) reported that 53.29% of his subjects used their monolingual English dictionaries in reading activities with a frequency of once a week. Hartmann (1983b:198) pointed out that over 83% of his predominantly upper secondary school sample said they used their dictionaries for reading texts. In Kipfer's study (1985:51) it was reported that out of 144 subjects only 44 (30.5%) said they used dictionaries for reading. Béjoint (1981:216) showed that the majority (60%) of his subjects used their monolingual dictionaries for mainly decoding in the written medium. As far as the nursing students at the University of Jordan are concerned, it can be concluded that reading comprehension represents a top priority, particularly reading textbooks, and the dictionary is very much relied upon by a majority of students at all levels.

5.5.4.1.2 Listening Comprehension

To understand spoken English in academic contexts by non-native students is more demanding than the task of reading comprehension. This is because of the nature of the process involved in listening comprehension. Rixon (1981:68) postulates that an efficient listener applies the following strategies:

1. Sorting out the purpose of listening, i.e., what does he want to know?
2. Predicting some of the information to be included in the utterance.
3. Assessing how much of that information is new and how much is already known.
4. Deciding how much of the message is likely to be relevant to the purpose of listening.

Rixon also points out that familiarity, or lack of it, with the subject matter and the language in question are factors that influence the use of these strategies.

As in reading comprehension, guessing meanings of unfamiliar vocabulary with the help of contextual clues has a significant role to play in listening comprehension; especially as the use of dictionaries is not often an easy option. But unlike in the context of reading comprehension, the listener has more opportunities to exploit the wealth of contextual aids available. Learning in subject matter situations is more contextualised than in ESP classrooms. In the technical classroom, much of the presentation mode is audio-visual. "The teacher does not just give a lecture **about** a machine, he **uses** the actual machine or a model of it as an integral part of the delivery..." (Hutchinson and Waters 1980:10).

On the other hand, the mode of presentation in technical classes employs extensive use of reference which is usually accompanied by gestures, actions, and even colloquial phrases. Because the technical teacher uses reference to other areas of knowledge in order to present a new concept, the overall knowledge of the world on the part of the student becomes an important factor in comprehension. Moreover, the student has to cope with modes of graphic information (charts, diagrams, graphs etc.) in order to be able to understand the spoken content delivered in the foreign language (ibid.:11 and 17).

As I did with reading comprehension, I shall present the findings on students' perceived needs in contexts of listening comprehension and relate these needs to the students' use of dictionaries in each of the listening contexts. But first let us consider the findings on the students' general need for help in practising listening to spoken English on medical/nursing subjects. A great majority of the students (92.8%, N=376) said they needed help in this respect; more than a half of these students (52.8%, N=198) reported 'much' need. In comparison to the need for help in reading comprehension, the need for help in listening comprehension seems to be more emphasised by the students as a greater number of students at all levels said they needed help in this area.

5.5.4.1.2.1 Lectures

Lectures form a major medium of learning in university education, and much of the students' and staff time is spent in lectures. In the Faculty of Nursing, both Jordanian and expatriate staff use English as a primary medium of teaching. Therefore, there is a felt need by both students and staff that students ought to be taught how to understand lectures in English efficiently. The need for help with the spoken English used in lectures seems to be apparent among all groups, particularly among the first-year students. More than 80% of the students (N=335) reported that they needed help in understanding lectures; among these students at least 40% wanted 'much' help. The nursing teachers also confirmed these results, as 19 out of 24 (79.1%) said the need in this area was 'high' and 5 (20.8%) rated this need as 'average'.

Nonetheless, it seems that the majority of students in the Nursing Faculty do not resort to dictionaries in the lecture context for comprehension purposes. The results indicate that only a few students reported using dictionaries to understand spoken English used in lectures. Of the sample, 76.5% said they 'never used a dictionary in a lecture context, and 81% of these students who said they did so reported a 'low' frequency. The results of the students' interviews showed a similar result, as 40 out of 41 students said they did not use dictionaries in understanding lectures. It was surprising, therefore, to find out that 17 nursing teachers out of 24 (70.8%) said that it was 'very important' for their students to use a dictionary in the context of lecture comprehension.

Although it is officially stated that English is the medium of instruction in the Faculty of Nursing, in actual practice this does not apply to all contexts of teaching. I attended a few lectures in the Faculty of Nursing to familiarise myself with the use of English in this context. Although the experience was fairly limited, I found that the Jordanian staff tended to use a good deal of Arabic and in many cases the students were dictated the English content of the lecture. What also explains the lack of dictionary use in lectures is that it is awkward, if not impossible, to look up words in a dictionary during the process of listening.

5.5.4.1.2.2 Seminars

The seminar context is a common communicative situation where most of the interaction is carried out through English. Staff and students regularly listen to an oral presentation and discuss its content as part of the syllabus. These seminars are usually held in the University teaching hospital and are based on actual cases drawn from the hospital itself. First-year students do not attend such seminars and therefore I exclude them from the discussion of dictionary use in this context.

The need for help in the context of seminars seems to be similar to that indicated in the context of lectures if we take the sample of students as a whole. Out of 278 students (the total number of the second, third and fourth-year levels), 238 (85.6%) expressed their needs for help, among these students 36.9% wanted 'much' help. However, there are variations between one level of students and another; even for one and the same level of students, the perception of needs seems to vary from one context to another.

Unlike the case of lectures, a considerable ratio of students (62.5%, N=174) reported using dictionaries in the seminar situation. The percentage of students who said that they were using dictionaries increases as the students move up the academic level, e.g., the highest percentage of dictionary users is among the fourth-year students. The frequency of dictionary use seems also to increase among the higher levels of students, i.e., third- and fourth-year students. Although the students reported using dictionaries in the seminar context, the frequency of use appears to be low in general and varies across the groups represented; more than 40% of the students who reported using dictionaries indicated a 'rarely' frequency. The interviews with the students showed that almost half of the sample (48.7%, N=20) said they used dictionaries in seminars, but it was noted that most of these students reported doing so with a very low frequency. On the other hand, 18 nursing teachers (75%) reported that they thought it was 'very important' for their students to use a dictionary in a seminar situation.

During the few seminars I was able to attend, I noticed that although more English seemed to be used, Arabic was to some extent still tolerated despite the fact that the students were always encouraged to communicate in English. This observation may explain partly why dictionaries seem to be more popular in seminars than in lectures.

5.5.4.1.2.3 Clinical Training

The University teaching hospital is located on the main campus and is very close to the Faculty of Nursing. Both students and staff spend a considerable amount of time in the hospital for training purposes. Almost all communication between staff members and students takes place in English, therefore I decided to explore how much need for help in English was required and to what extent dictionaries were used in this context. There is an obvious need amongst the students of all levels (except the first-year ones who do not attend clinical classes at this stage) to understand spoken English during clinical training. Out of 278 students, 87% (N=242) said they needed help; more than a third of them (39.6%) wanted 'much' help. The need seems to vary from one level to another. The nursing teachers confirmed these results, as 15 out of 23 (65.2%) said the need in this area was 'high' and 7 (30.4%) reported an 'average' level of need in the same context.

As far as dictionary use in this context is concerned, a majority of students (81.3%, N=226), excluding the first-year group, said they used dictionaries in understanding spoken English in clinical training rounds; 46% of these students reported a 'high' frequency of doing so. In terms of frequency, it is noticed that dictionaries are used more frequently in this context than in other contexts such as seminars. But the figures show variations from one group to another. The interviews with the students revealed that 21 students out of 41 (51.2%) said they were using dictionaries regularly in the context of clinical rounds. On the other hand, the nursing staff questionnaire showed that 62.5% (N=15 out of 24) said they thought it was 'very important' for their students to use a dictionary for the purposes of comprehension during clinical training. In fact, I was told by students that students attending clinical training were recommended a particular monolingual English dictionary, namely BAILIERE'S NURSES' DICTIONARY by Kasner and Tindall (pocket size), London: Baillière Tindall, 1912, to be used in this context. Much of the communication in this context is formal and usually takes place at the patients' bedsides. From the results above it seems that dictionary use is connected to the type of context/activity the students are engaged in and to the extent to which English (or Arabic) is being used. In the skill of listening comprehension, we have seen that dictionary use varied from one context to another and to some extent from one level of students to another. The conclusion that can be drawn from this is that dictionary use should be linked to the specific context of communication in which English is being put to use. The observation that dictionary use varies among the component groups in the sample lends support to this conclusion. Had the results been reported on the basis of the total sample alone, a false picture might have emerged as far as the use of dictionaries in relation to needs is concerned.

Furthermore, the finding that a considerable majority of the students indicated using a dictionary during clinical training fairly frequently calls for the attention of the ESP practitioner as well as the ESP lexicographer. In the reported literature of ESP, much of what has been done in the area of medical/nursing English has been based on written samples of language. The spoken English used in student nurse - staff nurse communication in the hospital setting has received little attention. For the ESP dictionary designer, the implication is that a pocket-type dictionary is expected to meet the lexical requirements of the students on the hospital floors. Consulting a large-size dictionary while standing at the patient's bedside can be cumbersome and embarrassing.

In the reported literature on dictionary use, it seems that using a dictionary to aid listening comprehension is less popular than to help in reading comprehension. For example, Béjoint (1981:216) reported that only 14% of the subjects used a dictionary for listening comprehension. Hartmann (1983b:198) indicated that fewer than 20% of his subjects reported dictionary use for listening comprehension. Iqbal (1987:82) reported only 8.7% of his sample using dictionaries in listening contexts. Although informative, these studies do not tell us much about the place of the dictionary in relation to the specific communicative activity the user is involved in, since all these studies confined their scope to the traditional classification of language skills, i.e., reading, listening, writing and speaking, and ignored those specific activities which call for the use of dictionaries.

5.5.4.2 Encoding Needs and Dictionary Use

In an academic situation where English is the major medium of instruction and communication, it becomes essential that learners are trained to communicate their intentions efficiently. Both ESP teaching and lexicography can play an important role in achieving this purpose. To do so, the dictionary designer and the syllabus writer need access to specific information on the domains of language use and the uses of dictionaries in such domains. In what follows I present and discuss the findings related to the needs among the students for help in encoding contexts and relate these to the findings on dictionary use in each context.

5.5.4.2.1 Writing

The writing skill is by no means a simple task for the non-native speaker of a target language; even for native speakers, writing can be demanding. Unlike speaking, writing can be difficult because of the absence of constant feedback from the receiver, and because the writer cannot use paralinguistic means which can help make the message understandable. In addition, academic writing is formal and therefore demands the use of a particular style and requires care in the choice of lexis. Difficulties in writing can take place at more than one level, i.e., lexical, syntactical, and rhetorical. Added to these levels is the conceptual level, that is, knowledge of the subject matter involved.

In this study, both students and nursing staff were invited to express their opinions about the felt needs for help in writing English. 374 students (92.3%) felt they needed help in writing or nursing topics; more than half of them (54%) reported 'much' need for help. In comparison to the results concerning the needs in reading comprehension, it is clear that more students at all four levels of students said they needed help in writing. Also the degree of need indicated seems to be higher in writing than in reading.

In order to find out the nursing teachers' attitudes towards the importance of writing skills for their students in contrast to reading skills they were asked to react to the statement: "Writing skills are less important for the students than reading skills." The results showed that 91.6% (N=22 out of 24) disagreed with the statement.

Having noticed punctuation problems in a sample of the first- and second-year students' written assignments, I decided to find out to what extent there was a need in this area. No less than 80% of the sample reported that they needed assistance in punctuation. On the other hand, out of 23, 60.8% (N=14) of the nursing teachers reported a 'high' degree of need in this respect, and 21.7% (N=5) reported an 'average' level of need whereas 3 teachers (13%) indicated a 'low' level of need.

An important area of writing is in the context of examinations. Because the amount and the type of writing involved in examinations depend on the type of questions employed, I first asked the nursing staff what types of questions they used and how often they used each. The results were as follows:

Table 7: Question types used by the nursing staff

Question Type	Used by (out of 22) N	%		Frequency low	average	high
yes/no (true/false)	18	81.8	N= %=	1 (5.5)	10 (55.5)	7 (38.8)
essay	17	77.2	N= %=	2 (11.7)	8 (47.0)	7 (41.1)
matching	17	77.2	N= %=	7 (41.1)	8 (47.0)	2 (11.7)
defining	22	100.0	N= %=	5 (22.7)	4 (18.1)	13 (59.0)
multiple choice	21	95.4	N= %=	6 (28.5)	3 (14.2)	12 (57.1)
	(out of 20)					
labelling	13	65.0	N= %=	4 (30.7)	7 (53.8)	2 (15.3)
other (mainly oral)	4	20.0	N= %=	- -	2 (50.0)	2 (50.0)

The question types that require writing are the essay and defining. Both seem to be used by the majority of teachers; therefore it can be said that it is not true that nursing students do not have to write in examinations, as is sometimes claimed.

The need in writing examination answers seems to be clearly apparent among all groups, particularly among the first-year level. 338 students (83.4%) reported their need for help in this context; more than 40% of them wanted 'much' help. The views of the nursing teachers are in line with these results. Out of 24 teachers, 20 (83.3%) said the level of need was 'high' and 3 (12.5%) reported an 'average' level of need.

Writing proficiency is also required in the context of note-taking from spoken sources. The results showed that more than 80% (N=332) said they needed help in this context; 38.8% of them reported 'much' need. Especially amongst the first-year and the second-year levels the need is evident. 23 nursing teachers (95.8%) said there was a need in this respect, of whom 16 (69.5%) reported 'high' level of need, and 6 (26.0%) said the level of need was 'average'.

As for dictionary use in writing activities, it seems that a majority of students at all levels recognise the need to refer to a dictionary for writing purposes. 63.9% of the students agreed with the statement: "Nursing students need to refer to a dictionary when they write in English", whereas 25.9% disagreed and 10.1% said they did not know. Although the general tendency suggests that students feel a need to use dictionaries in writing, there seem to be slight variations between one group of students and another according to the level of study.

The nursing students may refer to dictionaries while performing in one of the following contexts of writing: academic essays, technical reports, and nursing care plans.

5.5.4.2.1.1 Academic Essays

The academic essay has been described by an ESP practitioner as "a kind of game in which the writer, according to the extent to which he or she is familiar with the rules is able to use them, seeks to satisfy the demands of the reader/marker."

(Houghton 1984:47). Many ESP students are in trouble if they are not equipped with the necessary awareness of the rules of this 'game' and the skills in using them.

The students were asked about their needs for help in this context. The overwhelming majority of the students (91.3%, N=254 out of 278) reported that they needed help in essay writing; more than a half of them (54.7%, N=139 out of 254) indicating a high level of need. (The first-year students reported that they were not involved in essay writing during the first semester.)

The overall tendency seems to be consistent among all groups, but there are some variations between one level and another. In addition, out of the 24 nursing teachers 23 (95.8%) indicated that their students needed help in essay writing; of these 23 teachers, 20 (86.9%) reported a 'high' degree of need, while 2 (8.6%) reported an 'average' degree of need.

The overwhelming majority of the students in the three remaining levels (87.4%, N=243 out of 278) reported using dictionaries in writing their essays. The percentage of the students who claimed a 'high' frequency of use in this context was also high (62.1%, N=151 out of 243).

Third-year students seem to be the major users of dictionaries in writing essays, followed by fourth-year students. The students' interviews showed that 17 students out of 41 (41.4%) reported using dictionaries in writing essays with a 'high' frequency; all of the 24 teachers said that it was 'important' for their students to use dictionaries in the writing of essays; for two thirds of these teachers (66.6%, N=16) dictionary use in writing essays was rated as 'very important'.

5.5.4.2.1.2 Technical Reports

Reports in nursing are quite important as they are considered to be legal documents. Therefore accuracy is a must in this form of communication. The nurse on duty may need to write about the progress of the patient and describe changes in the patient's condition. It is important, therefore, that the nursing students are trained in writing reports as fluently and accurately as possible. Faulkner (1985:139), a nursing textbook writer, states that "Just as in verbal interactions, ambiguous words and vague phrases should be avoided when making written reports on patients".

In this survey, the students and the nursing staff were requested to indicate the extent to which there was a need for help in the context of specialist report writing. The results showed that 85.9% of the students, excluding the first-year students who said they did not write reports during the first semester, reported that they were in need of help, of whom almost half (46.4%) felt the level of need was 'high'. The reported needs seem to vary according to the level of study. It was also found that 23 out of the 24 nursing teachers (95.8%) said their students needed help in writing reports; 19 (82.6%) of these 23 teachers reported a 'high' degree of need in context while for 4 (17.4%) the level of need was rated as 'average'. A great majority of students (excluding the first year) (84.1%) said they used dictionaries when writing reports; over half of these students (54.7%) reported using dictionaries with a 'high' frequency. The results showed some variations between the constituent groups in the sample. For example, dictionary use seems to be more popular in this context among the second-year students than among the fourth-year students. In comparison to the context of essay writing, it seems that dictionary use in report writing is less frequent. The students' interviews revealed that 41.4% (N=17) reported using dictionaries with a 'high' frequency in the context of writing reports. On the other hand, 20 nursing teachers out of 23 (86.9%) said it was 'important' for their students to use dictionaries in the context of writing reports; out of these 20 teachers, 15 (75%) ranked dictionary use in this context as 'very important'.

5.5.4.2.1.3 Nursing Care Plans

An essential component in the nursing process is planning actions and procedures that the patient should receive from nurses. Planning care needs to be documented through special records known as care plans where each patient's progress is noted. Linguistically, care plans play a significant role in hospital communication. They have to be comprehensible not only to other nurses but to other members of the health team, and therefore accuracy and clarity should be maintained. According to Faulkner (1985:117) a care plan "should be readily understood by those who are expected to use it. As such it should be succinct and devoid of irrelevant material".

Out of 278 students 75.1% (N=209) said they needed help in this context; almost half of them (48.8%) said they needed 'much' help. Particularly amongst the second-year and the third-year students the need for help is clear. The lower percentage of fourth-year students is probably due to the fact that these students at this level are more proficient and more accustomed to writing care plans than the other levels of students. The students' needs for help in writing care plans was also reported by 23 nursing staff members (95.8%). Out of these teachers 20 (86.9%) said the level of need was 'high'.

Dictionary use in this context was reported by a majority of students (80.5%), excluding the first-year students who reported that they were not required to write care plans during the first semester of the academic year. The percentage of students claiming dictionary use is consistently high among the three groups of students but it decreases as the students move up the academic scale. Frequency of dictionary use is noticeable as almost half of the students (47.7%) reported a 'high' frequency in this activity. Frequency of use seems to be decreasing as the students move up the academic scale. The results also show that the second-year students are the main users of dictionaries in this context. The students' interviews show that 41.4% (N=17 out of 41) indicated using dictionaries while writing care plans with a 'high' frequency. Almost all of the nursing teachers

(N=23 out of 24) reported that it was 'important' for their students to use dictionaries in this context; 17 of these 23 teachers (73.9%) ranked dictionary use in this context as 'very important'.

It is significant that a majority of students in all years except the first reported using dictionaries in the various types of writing. This suggests that the students are in need of a production-oriented dictionary. Therefore, it is suggested that a sample of the English used in the students' written essays, reports and care plans should be incorporated in the linguistic database on which dictionaries will be compiled for productive purposes. Supporting this suggestion is a similar finding by El-Badry (in progress) which indicates that more than three-quarters of the subjects (78.8%) reported dictionary use in writing tasks. In Hartmann's study (1983b:198), it was reported that dictionary use in writing was reported by 74% of the subjects; and in Kipfer's survey (1987:45) it was found that dictionaries were used more for writing than reading (230 out of 292 (78.7%) reported dictionary use in writing, mainly short essays and reports).

It is impossible, it seems, to make universal generalisations about dictionary use as it is evident that different groups of users in different countries vary considerably in their use of dictionaries. This is possibly due to the multiplicity and complexity of variables involved, e.g., purpose of dictionary use, context of dictionary use, the types of dictionaries used, the type of users and their ages, levels of proficiency in the language. Indeed, variation in dictionary use seems to exist even within the same group of users as is clear in the results of this study.

In general, dictionaries so far seem to cater mainly for language readers rather than writers, and even the few learners' dictionaries claiming to meet the writer's needs have been shown by recent research not to be capable of this task as they try to meet a wide range of types of requirements by a wide range of users (cf. Nesi 1987).

5.5.4.2.2 Speaking

Encoding through the spoken medium is by no means a simple operation; nor is it a less important aspect of communication than other language skills. For ESP and lexicography purposes, it is important to know whether or not and to what extent students make use of dictionaries while engaged in speaking-related activities. Speaking is not only knowledge about language, it is a skill which cannot be developed without adequate training and practice (Bygate 1987:3). Involved in the skill of speaking, according to Bygate (ibid.:5), are motor-perceptive skills, i.e., "perceiving, recalling, and articulating in the correct order sounds and structures of the language", and interaction skills which involve "making decisions about communication, such as: what to say, how to say it, and whether to develop it, in accordance with one's intentions, while maintaining the desired relations with others".

At the University of Jordan, nursing students generally experience great difficulties in making themselves understood. One of the reasons for this problem is the examination system at the pre-university level. Because the G.S.E.C.E. does not have an oral component, both students and teachers usually relegate the objective of developing oral skills to a secondary position. Another reason is the lack of exposure that students have to natural spoken English. English is rarely spoken outside the classroom and even in the classroom some teachers use Arabic in their teaching of English. Out of the 405 students in the sample, 379 (93.5%) reported that they needed help in practising speaking about nursing topics; for two thirds the need was ranked as 'much'. In comparison to the needs in writing, both the percentage of students indicating needs and the percentage of students reporting 'much' need is higher in speaking than in writing. If we consider the four groups individually, it seems that the first-year students followed by the third-year students indicated a higher level of need than the second- and the fourth-year levels.

The students were also asked to report their needs for help in speaking against each of the main academic activities that might involve speaking.

5.5.4.2.2.1 <u>Lectures</u>

Although there are slight variations from one group to another, the overall tendency suggests that a vast majority of students (89.8%, N=364) expect help in this context. The nursing staff questionnaire, on the other hand, revealed that 22 teachers (91.6%) said their students needed help in speaking in lectures; half of these teachers ranked the level of need as 'high', and 8 (36.3%) ranked it as 'average'.

5.5.4.2.2.2 <u>Seminars</u>

Speaking in seminars was also reported to be an area of need by 88.4% of the total sample of students (N=278 as the first-year students were excluded), for half of them the level of need was ranked as 'much'. Compared to the needs for help in speaking in lecturers, both the percentage of students indicating needs and the percentage of students reporting 'much' need is lower in the case of seminars. Almost all of the nursing teachers (95.8%, N=23) reported that their students needed help in this respect; 15 of them (65.2%) indicated a 'high' level of need, whereas 6 (26%) stated that the level of need was 'average'.

5.5.4.2.2.3 <u>Clinical Training</u>

The nursing students also need training in speaking English during their clinical training rounds. 88.4% of the total sample (N=246 out of 278) reported that they needed help in this context, almost half of these students indicated a 'much' level of need.

Though the general tendency is similar amongst the various groups of the sample, there appear to be differences between them. The first-year students do not attend clinical training sessions in the hospital during the first semester and therefore they are excluded. It seems that it is mainly the second-year students who require more help than the other years. It was also found that all of the 24 nursing teachers in the sample said their students needed help in speaking during clinical rounds; 14 (58.3%) indicated that the level of need was 'high', whereas 8 (33.3%) reported that the level of need was 'average'.

5.5.4.2.2.4 Oral Examinations

Answering questions in oral examinations was also found to be an area of need. A great number of students (87.9%, N=356) said they needed help. More than a half of these students (53.1%) reported that they needed 'much' help in this context. This need is noticeable more particularly among the first-year students followed by the second-year students than in the case of the higher levels of students, i.e., the third and the fourth years. On the other hand, all of the nursing teachers in the sample said that their students needed help in the context of oral examinations; 62.5% (N=15) of these teachers reported a 'high' level of need in this respect.

5.5.4.2.2.5 Oral Presentations

A further area of spoken English in which help is required was reported to be oral presentations. A considerable majority of students in the sample as a whole (N=246 out of 278) and in the sub-samples reported that they needed help in this context. More than a half of the students reported 'much' need. The degree of need varied from one level of study to another, e.g., the degree of need is lower among the fourth-year students than among the second- or third-year levels. As for the nursing staff, it was found that 23 teachers (95.8%) said their students needed help: 16 (69.5%) ranked the level of need as 'high', and 6 (26%) reported an 'average' level of need.

As far as dictionary use in speaking contexts is concerned, the results of the students' questionnaire showed that preparing oral presentations was the only activity in which dictionaries were used for speaking purposes. The first-year students were excluded in considering the figures regarding dictionary use in this context as they all stated that this activity did not apply to their situation because they were not required to give oral presentations at that stage. Although three-quarters (N=209 out of 278) of the students' sample (75.1%) reported using dictionaries in this context, the percentage of students reporting 'high' frequency of use was found to be rather low (43.5%) in comparison with dictionary use in other contexts such as writing. The frequency figures vary among the sub-groups in the sample; for example, dictionary use in this respect seems to be more popular among the second-year students than among the fourth-year students. The results of the interviews with the 41 sample students showed that, excluding the first-year students who reported that they were not required to give oral presentations at the time of the study, 22 students out of 33 (66.6%) said they were using dictionaries while preparing oral presentations. The results of the interviews also revealed that only a small number of students (3 out of 41) reported using dictionaries in order to speak in other contexts such as lectures, seminars.

As regards the nursing teachers' attitudes, it was found that 22 out of 24 (91.6%) reported it was 'very important' for their students to use dictionaries in this context. While preparing their oral presentations the nursing students need access to semantic, syntactic, and phonetic information if their spoken performance is to be fluent and accurate. A dictionary that does not provide sufficient help in these areas is unlikely to be useful from an ESP teaching point of view.

Reporting on dictionary use for speaking purposes, Hartmann (1983b:198) reported that dictionary use for 'conversation' was indicated by 19% of his sample of users. Béjoint (1981:216) revealed that no more than 9% of his sample reported using dictionaries for 'oral composition'; he also found that dictionaries were more often used for written activities than for oral activities. Although it is hard to arrive at universal generalisations from a handful of studies carried out in different contexts with different orientations, there seems to be some evidence to suggest that dictionary use in speaking contexts is less popular among users than in other contexts. This is probably due to the nature of the speaking process which demands quick and spontaneous flow of language.

The dictionary reference situation seems to differ from one group to another; it would be difficult therefore to generalise the results. For example, in El-Badry's sample, only 27% were undergraduate university students, only 44% worked in the sciences, and only 51% were resident in Arab countries, and therefore the many different sub-groups in the sample may affect the results of the study. Moreover, the reported studies of dictionary use do not sub-classify the contexts of speaking and do not make a distinction between dictionary use in <u>preparing</u> for an oral activity and dictionary use in the speaking activity itself.

CHAPTER SIX
PURPOSES OF DICTIONARY USE IN RELATION TO LEXICAL NEEDS

6.1 Introduction

Having identified the communicative contexts in which dictionaries are used by the student sample, and having related these contexts to a profile of communicative needs, I shall now consider the categories of information the students consult in dictionaries. Nine types of such information were entered in the students's questionnaire; the students were requested to indicate whether or not, and with what degree of frequency, they used their dictionaries to look up each of these. One blank line was provided at the end of the list and the students were instructed to mention any other type of information they sought in dictionaries, and with what degree of frequency. The information categories included:

(1) meaning
(2) pronunciation
(3) prepositions
(4) other specific grammatical information
(5) spelling
(6) illustrative materials, e.g. plates, pictures, diagrams and drawings
(7) information in appendices
(8) syllabification
(9) etymology

This list was arrived at after consultation with a number of nursing staff and students before the questionnaire was finally administered. To make sure that the students understood what was meant by each item on the list, English equivalents and examples were provided in writing. In addition, linguistic and TEFL jargon was avoided in the wording of the items. For example, instead of asking the students whether or not they looked up <u>etymology</u>, they were asked to circle the frequency of looking up the origins of words/phrases of, e.g., Greek or Latin.

6.2 Lexical Needs of ESP Learners

Ideally, an ESP syllabus and/or an ESP dictionary should be designed in the light of the lexical needs of the target group of users. However, not much research has been reported in either the ESP and/or the lexicographical literature on the issue of ESP learners' lexical needs/problems. According to one authority (Swales 1985b:213) the issue of vocabulary in ESP has suffered serious neglect and "the situation still remains very unsatisfactory". If we, lexicographers and ESP practitioners, accept that it is our job to help learners of ESP, our products must be based on an understanding of the characteristics of the E in ESP. For example, both lexicography and ESP teaching should take into account that specialist English is not only written discourse, but also oral discourse. Any lexical corpus designed for dictionary and syllabus production should cover both written as well as spoken genres of English.

ESP practitioners seem to agree that it is semi-technical lexis that causes serious problems to ESP learners (cf. Barber 1962, Herbert 1965, Higgins 1967, Inman 1978, McDonough 1984, Kennedy & Bolitho 1984, Trimble 1985). The term 'sub-technical vocabulary' has been defined by Cowan (in Trimble 1985:129) as "context-independent words which occur with high frequency across disciplines". Trimble (1985:129) argues that the term covers "common words that occur with special meanings in specific scientific and technical fields". According to Kennedy and Bolitho (1984:57), sub-technical vocabulary "consists of those words which are not specific to a subject speciality but which occur regularly in scientific and technical texts, e.g., reflection, intense, accumulate, tendency, isolate and dense". Sub-technical lexis is a major constituent of scientific and technical language. Inman (1978:248) argues that excluding sub-technical vocabulary can mean excluding up to 80% of a text.

Another area of problematic ESP lexis is that of noun compounds, a characteristic feature of scientific and technical discourse. Trimble (1985:130) defines this phenomenon as "two or more nouns plus necessary adjectives (and less often verbs and adverbs) that together make up a single concept; that is, the total expresses a 'single noun' idea". Unfortunately, both sub-technical vocabulary and noun compounds seem to have been insufficiently treated in currently available dictionaries. According to Trimble (1985:128):

> "Non-native students faced with a confusing sub-technical vocabulary or with unfamiliar noun strings naturally went to their bilingual and/or 'standard' English dictionaries, neither of which were designed to give them much help with these particular problems."

It should be borne in mind that a scientific or a technical text is not made only of one level of lexis but of vocabulary items that range from the very technical to the very general. An ESP student is likely to look up any word that s/he finds difficult regardless of whether it is technical or sub-technical. Indeed, the division of specialist lexis into levels of technicality seems to be a matter of convenience to researchers and of no concern to students. The notion of sub-technical and of technical vocabulary is, moreover, not always clear-cut even among applied linguists. According to Baker (1985:91) "'subtechnical' as a category has proved to be elusive and confusing for many teachers, the reason being that the term has neither been clearly nor consistently defined in the literature".

The trend in the ESP literature has been that it is not the job of the ESP teacher to deal with technical vocabulary on the grounds that (1) such items are beyond the capacity of the ESP teacher and (2) such items are already taught as part of the subject matter courses. But in practice, this division of responsibilities is not always followed and/or appreciated. According to McDonough (1984:103), "specialist vocabulary is often a no-man's-land, perceived by subject-teachers to be part of language teaching, and by language teachers to be outside their competence". In a team-based collaborative approach to both ESP teaching and ESP lexicography where ESP teachers, subject teachers, pedagogical lexicographers and publishers are represented, such problems should not occur. The selection of lexical content to be entered in an ESP syllabus or an ESP dictionary should not be based on the notion of technicality, but rather on the notions of difficulty and usefulness for the particular group of learners.

6.3 Semantic Needs and Dictionary Use

Accessing semantic information is a major concern for both ESP learners and ESP teachers. Whether engaged in language use productively (writing and speaking) or receptively (listening and reading), lack of sufficient vocabulary and/or inability to use it appropriately can cause serious problems of communication. In this study, I asked the students and their nursing teachers, through the questionnaires, about their perceived needs for help with the meanings of words. The results showed that the great majority of students (N=376 out of 405, 92.8%) expressed a need for help with meaning; more than 44% of them wanted 'much' help. This need was also reported by the nursing staff. Almost all members of staff (N=23) said their students needed help in the area of meanings of 'general' vocabulary; 20 (86.9%) of them said the degree of need was 'high'. As for the area of medical terminology, it was found that although almost all of the staff members (N=23) reported that their students needed help, only 13 (56.5%) reported a 'high' level of need, whereas 8 (34.7%) ranked the need as 'average'. This suggests that the students face problems with both types of vocabulary, general and technical, and that priority should be given to 'general' vocabulary.

Not only was the number of students who reported using dictionaries for meaning found to be large (392 out of 405) , but also the percentage of those reporting a 'high' frequency was very high (73.4%). In the follow-up interviews with 41 students, further data were obtained in this respect. Almost the whole sample (N=40) said 'yes' when asked whether they looked up the meanings of words of both types, technical and general. On the other hand, the nursing staff (N=24) said it was important for their students to use dictionaries to look up meanings of words; 22 of these teachers (91.6%) ranked dictionary use for meaning as 'very important'.

Other researchers have also reported that looking up semantic information in dictionaries was a major priority among dictionary users. El-Badry (in progress) found that 94.8% of her sample stated that they searched for meaning in their dictionaries; for 64.9% of El-Badry's subjects, looking up meaning represented at least 50% of total dictionary look-up operations. Iqbal (1987:79) reported that "looking up word meaning (99.43%) in the dictionary is the most basic requirement..."; as for frequency of use, Iqbal (1987:83) mentioned that 39.14% said they looked up meanings once a week. Looking up word meaning also came top in a list of eight information classes consulted in dictionaries used by Hartmann's (1983b:198) sample. Furthermore, 87% of Béjoint's (1981:215) subjects "placed meaning among the three most often sought-after pieces of information".

ESP students not only resort to dictionaries but, in many case, they also seek information on word meaning from alternative formal and informal sources. Among such informal sources are fellow students. I asked the students, in both the questionnaire and the interviews, whether or not (and how often) they asked each other for help on word meaning. A very high percentage of students (93%, N=377 out of 405) reported using this strategy of lexical acquisition; almost half (47.4%) of the students said they did so with a 'high' frequency. The follow-up interviews confirmed this finding: 37 out of 41 students (90.2%) said they regularly asked their fellow students for meanings of technical terms. In addition, 32 (78%) reported they asked their fellow students for meanings of general vocabulary. It is natural, and more convenient, that students resort to each other for help on word meaning, as fellow students are almost always accessible and communicating with them is informal. It remains doubtful, however, whether such sources of lexical information are reliable and/or accurate.

The nursing teachers also act as meaning suppliers. The nursing staff questionnaire showed that all respondents (N=24) said they explained the meanings of words when they were teaching; 11 of these teachers (45.8%) said they did so with a 'high' frequency, whereas 11 (45.8%) reported an 'average' frequency. The students' questionnaire and the follow-up interviews showed that a majority of students also depended on their nursing teachers as far as meaning was concerned. Across all levels and groups in the sample, it was found that most students (368 out of 405) said they used this strategy, though with varying frequencies. Taking the total sample, it is interesting to find that the percentage of students who claimed to ask nursing staff for meaning with a 'high' frequency (38.3%) is lower than the percentage of students who reported asking fellow students for the same purpose (47.4%). The results of the follow-up interviews showed that 30 out of 41 students (73.1%) said 'yes' when asked whether or not they asked their nursing teachers for meanings of 'technical' words. However, only 13 (31.7%) reported asking their nursing teachers for meanings of 'general' vocabulary.

ESP teachers also have a role in providing semantic information. Students usually expect their language teachers to provide meanings of unfamiliar vocabulary. To assess the role of the ESP staff in this respect, I asked the ESP teachers whether or not they explained meanings of words to students and the students whether or not they asked their ESP teachers for word meaning. The interviews with the ESP staff revealed that all the 6 members of staff said they usually explained word meaning whenever students asked and whenever they thought a particular item needed to be explained. In addition, they all indicated that they were asked by students for meanings of unfamiliar lexis. On the question of what type of vocabulary items they were usually asked about and/or they explained, I was told that it was both 'technical' and 'general' words. The ESP teachers also drew my attention to their observation that, particularly during their first two years of study, the students mainly asked about meanings of medical terms. The teachers also confirmed that their students had problems with 'sub-technical' lexis, the general words which have very specialised meanings in science.

A great majority of students (87.4%) claimed to be seeking vocabulary meaning from their ESP staff; more than a third of them reported doing so with a 'high' frequency. The sub-samples seem to vary in terms of frequency, but the overall tendency seems to be that the 'high' frequency percentages are consistently higher than the 'average' or the 'low' frequency percentages. On the other hand, it was found that 24 (58.5%) of the students interviewed said they asked their ESP teachers for the meanings of 'technical' vocabulary, and that the same percentage (58.5%) reported asking their ESP teachers for the meanings of 'general' vocabulary items.

A further source of semantic information usually available to ESP students is glossaries in subject-matter or ESP textbooks. These are lexicographical products and therefore need to be taken into account in considering sources of lexical information and the students' strategies in accessing such information. We know little about the learners' use of textbook glossaries, and therefore I asked the students whether or not, and if so, to what extent they made use of them. More than three quarters (77.7%) of the students claimed reference to meaning in textbook glossaries, and more than a third (39.6%, N=125 out of 315) of them reported doing so with a 'high' frequency. The users of this strategy seem to be mainly fourth- and third-year students. To what extent glossaries are useful/intelligible remains an area for further research. Almost all the subject textbooks used in the Faculty of Nursing were written for native speakers of English (mainly British and American), and therefore it is difficult to accept that ESP students have no difficulties in using them.

We must also add to the strategies mentioned above for accessing meaning that of guessing from context which I discussed within the context of reading (Chapter 5, Section

5.5.4.1.1): 95% of the total sample claimed to use guessing from context and 66% of them reported a 'high' frequency. If we group all these strategies of lexical reference together, and concentrate on the total sample (i.e., out of 405) percentage of users and the 'high' frequency data, we get the following profile:

Table 8: Students' strategies in accessing semantic information

Strategy	% of users	% of 'high' frequency users
Dictionary use	96.8	73.4
Guessing from context	95.5	66.6
Asking fellow students	93.0	47.4
Asking nursing staff	90.8	38.3
Asking ESP staff	87.4	38.1
Reference to textbook glossaries	77.7	39.6

In comparison to other strategies, dictionary use thus seems to be the most widely relied upon source of semantic information. This is probably because the dictionary is the most accessible and available to every student. Guessing may not be accurate, asking fellow students not reliable, and asking teachers not easy, therefore the dictionary seems to be the most handy and accurate reference source for students. The finding that both ESP staff and subject staff are expected to provide semantic information and that both sides explain meanings of words, both technical and general, implies that it is not only subject staff who are to be expected solely to be responsible for providing meaning. The ESP teacher will willy-nilly be asked questions by students on semantic matters and therefore must be prepared to play a role in this area.

During the exploratory period of my field work, some students asked me whether I could recommend a dictionary of technical abbreviations since they often faced some problems, especially when they listened to lectures or when writing or taking notes. This motivated me to explore their needs in this area through the questionnaire, the interviews and the staff questionnaire. A great majority of students (89.1%) expressed their need for help with abbreviations: no less than 50% of the students in each group reported 'much' need. The students also expressed their concern with abbreviations during the interviews: 31 students out of 39 (79.4%) said they needed help, among these 31 students, 20 (64.5%) reported 'much' need and 6 (19.3%) reported an 'average' level of need. The nursing teachers shared the opinion that students needed help with abbreviations: almost all the respondents (N=23) said their students needed help; 13 (56.5%) reported a 'high' level of need, and 8 (34.7%) said the level of need was 'average'.

What type of lexis do ESP learners regularly look up? First, I think the answer to this question depends on factors such as the level of the learner (i.e., fourth year vs. first year), the proficiency level in English (i.e., advanced vs. intermediate), and the type of text the student is dealing with (i.e., anatomy vs. psychology). There is evidence in this study that the nursing students, at all levels but in varying degrees, look up words of both a general and technical nature. The Dictionary Using Diaries also showed that the first-year students looked up words which can be described as technical, sub-technical and general. If we consider the lists of words looked up by such students we notice that those looked up can be classified as:

1. technical/subject specific, e.g., <u>hydrolytic</u>, <u>vesicles</u>, <u>lysis</u>, <u>mesenchyme</u>, <u>acinus</u>, <u>erythroblast</u>, <u>karyolymph</u>;
2. sub-technical/general words but with a specific meaning in nursing/medicine, e.g., <u>migrate</u>, <u>borders</u>, <u>contract</u>, <u>terminal</u>, <u>air-ways</u>, <u>labour</u>, <u>discharge</u>, <u>cell</u>, <u>conduct</u>;
3. general academic vocabulary, e.g., <u>pertains</u>, <u>process</u>, <u>implication</u>, <u>conjugate</u>, <u>tendency</u>, <u>synthesis</u>, <u>constitute</u>, <u>concept</u>, <u>contrast</u>, <u>characteristics</u>, <u>relatively</u>, <u>successive</u>.

We also notice that the following types of words were looked up:

1. nouns and noun compounds, mainly names of parts of the human body such as pollex, girdles, myofibrils, tympanic membrane, oesophagus, ovary;
2. nouns that are names of processes such as mitesis, degeneration, absorption, metabolism, transfusion, digestion;
3. verbs that describe certain notions such as a method or process, e.g., enter, eject, contract, withdraw, insert, inflate, connect, diffuse, diminish, combine, exchange;
4. adjectives describing certain notions like shape, size, colour, e.g., lobulated, elongated, dense, membraneous, bluish, flask-shaped, squamous, ciliated, colloidal;
5. adjectives describing directions/locations, e.g., terminal, perpendicular, superficial, postal, exterior, axial, horizontal, internal, inverted, parallel;
6. adverbs describing degree, e.g., approximately, slightly, sharply, considerably, significantly, relatively.

I also noticed that some students looked up general words such as definite, simultaneously, and action in a technical dictionary, whereas other students looked up technical words such as polyuria and ameoboid in a general dictionary. This suggests that some students do not distinguish between 'technical' and 'general' vocabulary, and that they are likely to look up an unfamiliar word in any dictionary.

6.4 Phonetic Needs and Dictionary Use

A basic requirement for ESP learners is to be able to understand, and to produce intelligibly spoken English which is used in academic contexts, e.g., lectures, seminars. This implies that the learner has to master sufficiently the English phonological (stress and intonation) system. This is a difficult task for a non-native learner of ESP; even for the native speaker of English, the pronunciation of medical terminology can be difficult. Gimson (1981:252) argues:

> "There is perhaps room for a pronouncing dictionary which omits all common words and deals only with the ever-increasing number of technical words, for the pronunciation of which even native speakers often need and seek guidance."

I obtained empirical data on the students' needs for help with pronunciation and on their strategies in acquiring it. Both the nursing staff and the ESP staff gave information on the needs of students for help with pronunciation. These needs were supplemented with data on the students' use of dictionaries. A great majority of students (93%) said they wanted help with pronunciation; 41% of them reported they needed 'much' help. First-year students particularly seem to require more attention. The results of the interviews similarly showed that 35 students out of 39 (89.7%) said they needed help with pronunciation; almost half of whom rated their needs as 'much' and nearly one third (31.4%), N=11) rated their needs as 'average'. As for the views of the staff, both the nursing and the ESP teachers, shared the opinion that their students needed help with pronunciation. The results of the nursing staff questionnaire showed that 22 respondents out of 23 (95.6%) said their students needed help with pronunciation of general vocabulary; 15 of these 22 students (68.1%) reported a 'high' degree of need. The need for help with pronouncing medical terminology was reported by 22 respondents (out of 23) (95.6%), out of whom 15 (68.1%) ranked the level of need as 'high'. The nursing teachers were also asked whether they taught their students how to pronounce words as part of their teaching and to what extent. The results indicated that all of them said they did so, 11 (45.8%) with a 'high' frequency, and 7 (29.1%) with an 'average' frequency.

All of the ESP teachers interviewed reported that they were often asked by their students, especially at the first-year level, to help with pronunciation of words of both types, technical and general. All of them also added that they helped their students with pronunciation, particularly by correcting wrong pronunciation. One of these teachers admitted that he deliberately avoided helping students in pronouncing medical terms because he himself felt uncertain of the correct pronunciation.

Having noticed some students transliterating on their textbooks and notebooks in Arabic, I decided to explore to what extent this strategy was being used by the students in general. A question was entered in the questionnaire and in the interviews, and examples were provided to explain what was meant by the question. This strategy does not seem to be popular among the students in general: more than 60% said they never used it, and almost half (44%) of those who said they used it reported a 'low' frequency of use. The use of this strategy also seems to vary considerably from one level of student to another; for example, fewer fourth-year students than first-year ones said they resorted to transliteration. The follow-up interviews also revealed that only 13 out of 39 (33.3%) students said they transliterated words they found difficult to pronounce.

To shed more light on the students' strategies of acquiring phonetic information and on the possible sources of such information, I compared the students' use of dictionaries for pronunciation with other reference strategies. Just over two thirds of the students (68.4%) reported consulting dictionaries for pronunciation, and only a third (33.5%, N=93 out of 277) reported a 'high' frequency. The first-year students turned out to be the main users of dictionaries for this purpose, which suggests that they are in much more need of phonetic help than higher proficiency students. The relatively lower percentage of users in comparison to dictionary use for meaning can be explained by reference to the type and intelligibility of phonetic indication in dictionaries, and to the students' own awareness and proficiency in accessing such information. The follow-up interviews, on the other hand, showed that 36.5% of the students said they consulted dictionaries for pronunciation of 'technical' terminology and 29.2% of the students reported using dictionaries for pronunciation of 'general' words. In addition, the nursing staff questionnaire showed that all the respondents claimed it was important for their students to look up pronunciation in dictionaries; 17 of them (70.8%) ranked the degree of importance as 'very important'.

Often students seem to rely on other sources when they need phonetic information. The vast majority (87.4%, N=354 out of 405) reported asking their fellow students when they needed to know how a word was pronounced. About one third (34.7%, N=123 out of 354) reported a 'high' frequency of using this strategy and a similar percentage reported an 'average' frequency. The use of this strategy is particularly apparent among the third-year students. In addition, 60.9% of the students interviewed said they consulted their fellow students for the pronunciation of 'technical' vocabulary. The results of the interviews also showed that almost half (48.7%) said they consulted their fellow students about the pronunciation of 'general' vocabulary.

Similarly, a high percentage of students (83.7%, N=339 out of 405) said they asked their ESP teachers for advice on pronunciation, more than a third of them (N=119 out of 339) reported a 'high' frequency. The results of the follow-up interviews showed that 41.4% of the students said they asked their ESP staff for help with pronunciation of 'technical' vocabulary and that 35% consulted their ESP teachers on pronunciation of 'general' vocabulary.

Asking the nursing staff about pronunciation was reported by a majority of students (79%, N=320 out of 405) with less than a third (23.1%) of them reporting a 'high' frequency. It is interesting to notice here that whereas for meaning, more students said they asked their subject teachers more frequently than asking their ESP teachers, for pronunciation the opposite was noticed. This may reflect something of the students' expectations in that they consider the subject staff more reliable for semantic information, whereas in pronunciation the ESP staff may be expected to be more proficient. The third-year students seem to be in the lead in that both the percentage of users as well as the frequency of consultations are higher than for other levels of students. Furthermore, 82% of the students interviewed reported consulting their nursing staff about the pronunciation of 'technical' vocabulary and 70% of the sample said they asked their nursing teachers about pronunciation of 'general' vocabulary.

Taking the total sample of students and the 'high' frequency data, the strategies of acquiring phonetic information can be ranked as follows:

Table 9: Students strategies in accessing phonetic information

Strategy	% of users	% of 'high' frequency users
Asking fellow students	87.4	34.7
Asking ESP staff	83.7	35.1
Asking nursing staff	79.0	23.1
Consulting dictionaries	68.4	33.5
Transliterating	37.0	16.0

Therefore, unlike meaning, pronunciation seems to be sought first from non-dictionary sources. This is probably due either to the insufficient treatment of pronunciation in existing dictionaries or to the students' unfamiliarity with the phonetic systems as used in dictionaries. The findings also suggest that students seek and expect help in the area of pronunciation from both ESP staff and subject staff. The results also consistently show that students wanted help with the pronunciation of both types of vocabulary, technical and general.

Other researchers have also reported on the use of dictionaries for pronunciation by their subjects. El-Badry (in progress) found that 69.1% of her subjects said they used their dictionaries to look up pronunciation. Iqbal (1987:169) discovered that looking up pronunciation did not figure prominently among his subjects as only 20.14% said they look up pronunciation once a week, and 16.14%, once a term. Hartmann (1983b:198) similarly reported that pronunciation seemed relatively unimportant for his subjects as looking up pronunciation came in the sixth position among eight information classes. In Béjoint's study (1981:215), only 25% of the students placed pronunciation among the three most often sought-after pieces of information. However, it is hard to compare the findings of this study to other studies because of the peculiarities of each sample of users and the multiplicity of factors involved in each situation. Yet, it remains an important observation that pronunciation is looked up less frequently and by fewer users than meaning. This does not, however, underestimate the importance of pronunciation for particular groups of users such as ESP learners.

6.5 Syntactic Needs and Dictionary Use

As ESP learners are expected to engage in acts of productive language use, they need to know not only what words mean, but also how to use them appropriately. This implies that they are required to be aware of the syntactic behaviour of words as part of their overall lexical proficiency. Being pedagogical tools, dictionaries are expected to play a role in this respect. Such a role is best achieved through due reference to the latest linguistic models of describing the language/s to be represented in the dictionary.

Although it is often noticed that the nursing students join the ESP programme with a reasonable knowledge of grammar in terms of being able to manipulate particular grammatical forms, e.g., changing the active voice into passive voice, they usually cannot express themselves correctly. This is because their proficiency in grammar is passive and because much of the pre-ESP classroom input usually concentrates on the sentence level rather than the discourse level. I asked the students and their nursing teachers whether they felt there was a need for help with grammar, and if so, to what extent. The results of the students' questionnaire showed that the majority (89.8%, N=364 out of 405) indicated their need for help; 45.3% (N=165 out of 364) of them wanted 'much' help, and 31% (N=113 out of 364) wanted 'average' help in the area of grammar. Out of 39 students interviewed, 30 (76.9%) said they needed help in grammar, among these 30 students 19 (63.3%) reported 'much' need whereas 7 (23.3%) reported an 'average' need. These results were complemented by similar results obtained via the nursing staff questionnaire. Almost all the respondents (N=23, %=95.8) agreed that their students needed help in using grammatically correct sentences when speaking: 16 (69.5%) reported a 'high' level of need and 6 (26%) an 'average' level. In addition, 23 of the teachers (95.8%) said their students needed help in using grammatically correct sentences when writing; 17 (73.9%) reported a 'high' level of need and 6 (26%) said their level of need was 'average'.

As far as dictionary use was concerned, the students' questionnaire revealed that although 60.9% (247 out of 45) of the sample reported looking up prepositions in dictionaries, only 19% of these 247 students indicated a 'high' frequency, whereas 42.1% reported a 'low' frequency. It is mainly the first-year students who seem to look up prepositions in dictionaries. 91.6% of the nursing teachers (N=22) said it was important for their students to look up prepositions in dictionaries; 13 out of the 22 (59.1%) rated looking up prepositions as 'very important'. My own experience suggests that students tend to make errors in their choice of preposition when speaking and/or writing. Textbooks in the fields of medicine and nursing, particularly anatomy and physiology, demonstrate frequent use of prepositional phrases. For example, consider these sentences from <u>Structure and Function in Man</u> (1982) by Jacob et al., a textbook used by the first-year students:

1. An organ is <u>composed of</u> cells <u>integrated into</u> tissues serving a common function. (p.11).
2. A substance closely resembling the structure of hydroxyapatite [Ca3 (po4)2]3. Ca (OH)2 <u>makes up</u> the major portion of salts presented in bone. (p.107).

Therefore, dictionaries for ESP learners should enter such verbs and prepositions and provide guidance on how to use them.

The students were also asked whether or not, and if so how often, they looked up other specific grammatical information, such as whether a verb was transitive or intransitive. Although more than half of the sample (55.5%, N=225 out of 405) claimed that they consulted specific grammatical information, almost half (46.6%) of these 225 students reported a 'low' frequency. Similarly, the results of the follow-up interviews showed that only 12.2% of the sample claimed to consult grammatical information with a 'high' frequency. As for the attitudes of the nursing staff, it was found that 20 teachers out of 24 (83.3%) said it was 'important' for their students to look up how words can be used grammatically; among these 20 teachers 12 (60%) described this as 'very important'.

To compare the students' use of dictionaries for checking grammatical information with other strategies, I asked the students through the follow-up interviews whether or not they consulted their fellow students, their nursing teachers, and their ESP teachers for grammatical information. The results showed that only 19.5% of the students interviewed said they asked their colleagues, only 2.4% said they asked their nursing teachers, whereas 36.5% reported asking their ESP teachers about grammatical issues.

In general, not many students seem to look up grammatical information in dictionaries. This finding may be explained by reference to one or more of the following reasons:

1. The nursing staff do not in fact emphasise grammatical accuracy; errors are usually tolerated; it is considered more important to "get the subject content right", as one of the nursing teachers put it.
2. The students are generally not aware of the fact that some reference books, especially learners' dictionaries, provide a wealth of grammatical guidance to help them with writing and speaking tasks.
3. The students do not generally understand grammatical terminology, e.g., the abbreviation v.i. for intransitive verbs.

In her study, El-Badry (in progress) discovered that consulting dictionaries for parts of speech was reported by 58.7% of the subjects. However, Iqbal (1987:137) found that 86.8% of the students reported consulting the dictionary on grammatical matters, e.g., word-classes. Looking up grammar came in second place, in terms of frequency, after meaning, in Hartmann's study (1983b:198). Similarly, in Béjoint's survey (1981:215) it was reported that looking up syntactic information came second after meaning as 53% of the students placed grammar among the three most often sought after pieces of information. By contrast, only 3 students out of the 292 subjects surveyed by Kipfer (1985:63) reported having used their dictionaries for looking up parts of speech in a writing context. Such conflicting results suggest that it is hard to generalise aspects of dictionary use from one situation to another, since each group of users has its own peculiarities and needs.

6.6 Orthographic Needs and Dictionary Use

Getting spelling right, especially in an area such as medical/nursing English, can be problematic even to native speakers of English. For ESP nursing students, accuracy in writing plays an important role in the communication between members of the nursing and medical teams. Despite this, very little research has been reported in the ESP literature on the problems of spelling among particular groups of ESP learners. However, of particular relevance to the Jordanian situation are two studies on spelling errors made by Jordanian university students. These are: Ibrahim (1978) and Abd-El-Jawad (1986).

Ibrahim (1978:208) argues that interference by Arabic spelling with English spelling is unlikely because of the difference in the scripts in the two languages. Jordanian undergraduates at the University of Jordan are reported to make mistakes in spelling because of the inconsistency and the non-phonetic nature of English spelling. Furthermore, the fact that Arabic has a different sound system from English affects students' spelling. Arab students naturally substitute [b] for [p] and [f] for [v] because Arabic lacks the sounds [p] and [v]. Hypercorrection is yet another reason for students mispellings. Language teachers tend to overcorrect students mistakes with [p] and [b] and this leads students to write *clup and *hapit.

Ibrahim also notes that because some learners perceive certain similarities between two or more words, they tend to spell one of these words on the pattern of the other/s. Analogy can be phonetic and therefore students produce *tought by analogy with bought. It can be orthographic and this accounts for errors like *maney and *mony. Grammatical analogy can occur and thus students write *cann't or *can not, *heared and *toled. On the other hand, word derivation in English is somewhat arbitrary, and therefore students misspell safety as *savety, knowing that the [v] does not change in words like <u>slavery</u>. Students also make errors when adding suffixes in order to form words, e.g., *pronounciation. Ibrahim suggests that the foreign learner has to be made aware of the rules and their exceptions so that they can avoid overgeneralisation, which itself is a further reason for spelling errors.

Similarly, Abd-El-Jawad (1986) analysed a corpus of spelling errors made by Jordanian undergraduates at the University of Yarmouk. Abd-El-Jawad blames the students' unawareness of spelling rules, in addition to the factors already put forward by Ibrahim (1978), e.g., analogy, differences between Arabic and English, and recommends that students should be encouraged to use their dictionaries as often as possible to check their spelling.

An overwhelming majority of the nursing students in this study (90.8%, N=368 out of 405) reported that they needed help in spelling, with almost half of them (49.1%, N=181 out of 368) rating the level of need as 'much'. Moreover, in the follow-up interviews, 33 students out of 39 (84.6%) said they needed help with spelling; 17 out of 33 (51.5%) reported a 'high' level of need, whereas 9 (27.2%) indicated an 'average' level. As far as the nursing staff were concerned, it was found that 22 teachers out of 23 (95.6%) said their students needed help with spelling, for 16 (72.7%) the level of need was said to be 'high'.

Insofar as the use of dictionaries to check spelling was concerned, the results showed that a large number of students in the sample (70.8%, N=287) claimed to be using their dictionaries for looking up spelling, but less than a third of these users

(29.9%, N=86 out of 287) reported a 'high' frequency. Furthermore, the results of the follow-up interviews showed that 46.3% of the sample (N=19 out of 41) said they looked up the spelling of 'technical' terms with a 'high' frequency, and that 43.9% (N=18 out of 41) reported looking up the spelling of 'general' vocabulary with a 'high' frequency.

All the nursing teachers surveyed (N=41) said it was 'important' for their students to use their dictionaries to look up spelling; 17 (70.8%) rated the degree of importance as 'very important'.

Not only are dictionaries an obvious source of information on spelling, but students also consult fellow students, nursing staff and ESP teachers. 63.4% of the students (N=26) interviewed said they consulted each other on matters of spelling 'technical' terms, in comparison to 53.6% (N=22) who said they asked their fellow students about spelling 'general' vocabulary. Asking nursing staff for the spelling of 'technical' terms was reported by 14 students (34.1%), whereas asking nursing staff for the spelling of 'general' words was reported by 29.2% of the sample (N=12). In contrast, 46.3% of the students (N=19) reported asking their ESP teachers for the spelling of 'technical' vocabulary, and 48.7% of the students (N=20) said they asked their ESP teachers for the spelling of 'general' vocabulary. This suggests that, in addition to dictionaries, students tend to resort first to each other when in need to check spelling, and then to their ESP teachers and nursing teachers respectively.

Other studies (Hartmann 1983b, Béjoint 1981, Kipfer 1985) confirm that spelling forms a major dictionary activity and that users of various backgrounds do need help in this respect. For the ESP syllabus designer, the implication is that spelling aids should be provided in the form of an integrated set of materials; for the ESP dictionary designer this means that provision must be made for spelling guidance.

6.7 Dictionary Search for Illustrative Materials

By illustrative materials, I mean the non-verbal dictionary information usually entered in the form of coloured as well as black and white plates, pictures, diagrams, or drawings of various sizes. A large number of students in the sample (85.4%, N=346 out of 405) claimed to have referred to such materials in the dictionaries they used. More than one third (37.5%, N=130 out of 346) of the students reported a 'high' frequency in addition to 41.9% (N=145 out of 346) who reported an 'average' frequency. But the results show some variation between the sub-samples. For example, more first- and second- year students than third- and fourth-year students reported a 'high' frequency. In addition, 70.7% of the students (N=29) interviewed said they referred to illustrations embodied in dictionaries they used. These findings imply that the students appreciate the inclusion of illustrations in dictionaries. As far as dictionary-making for ESP learners is concerned, the role of non-verbal illustrative materials, and if possible the use of colour, should not be underestimated. An appropriate picture or drawing can add to the clarity of the definitions not only of the names of objects but also of processes and procedures. Such materials are also likely to stimulate the students' curiosity and interest and thus making the operation of looking up interesting.

6.8 Dictionary Search for Information in Appendices

The students were also asked whether or not, and if so, how often they referred to appended information (such as tables or units of measurement) which are usually included at the end of the dictionary. Out of the total number of students surveyed, 73.8%, (N=299 out of 405) reported regular consultation of dictionary appendices, though with only 29.7% (N=89 out of 299) reporting 'high' frequency. The percentage of students who claimed reference to appendices was shown to be increasing as they move up the academic scale, although the reported frequency varies quite considerably from one level to another. The results of the follow-up interviews also showed that no less than 53.6% (N=22) of the sample claimed they regularly referred to appendices in the dictionaries they used. In principle, there is little wrong with appendices in dictionaries as long as such information is of direct relevance and potential help to student users. The appended material should not be restricted to the subject matter of nursing, but be of practical assistance to the process of language learning. ESP teachers as well as subject teachers ought to be involved in selecting such materials to be included as appendices in dictionaries for ESP learners.

6.9 Dictionary Search for Syllabification

The word 'syllabification' was used neither in the questionnaire nor in the follow-up interviews. Instead, students were asked whether or not, and if so, how often they used their dictionaries in order to find out how a word was sub-divided, and an example was provided. The results of the students' questionnaire, showed that just a little over one third of the sample (39.7%, N=161 out of 405) claimed to be using dictionaries for such a purpose. In addition, more than 60% (N=98 out of 161) of the students who said they looked up word-division reported a 'low' frequency; even those students who reported 'high' frequency were found to be few (11.1%, N=18 out of 161). A similar finding was also obtained in the follow-up interviews; 95.1% (N=39) of the sample reported they never used dictionaries to find out how words can be divided into syllables. It is interesting, however, to note that 52.1% (N=12 out of 23) of the nursing teachers reported that it was 'very important' for their students to use their dictionaries for this purpose. For pronunciation and spelling purposes, I think it is a reasonable lexicographical decision to hyphenate both headwords and their phonetic transcription to enable the learner to pronounce, and write, a word/term conveniently.

6.10 Dictionary Search for Etymology

Less than a third of the sample (32.3%, N=131 out of 405) indicated using dictionaries for information on word origin; more than half of these 131 students (58%, N=76) said they did so, but 'rarely'. Those who reported a 'high' frequency numbered only 18 out of 131 (13.7%). The follow-up interviews yielded a similar finding; out of the 41 students interviewed, only 2 reported looking up etymological information in dictionaries. The results of the nursing staff questionnaire showed that 30.4% (N=7 out of 23) of the respondents thought it was 'very important' for their students to use their dictionaries for information on word origins.

Other researchers have also found that etymology did not capture the interest of dictionary users. El-Badry (in progress) reported that only 1.6% of her sample said they consulted dictionaries for information on etymology. In Hartmann's survey (1983b:198), only 12% of the subjects were reported to have looked up etymology regularly, and etymology was ranked in the seventh postion in a list of eight categories of information looked up in dictionaries. Iqbal (1987:83) also stated that only 5.7% of his sample said they looked up etymology regularly, and etymology was ranked at the bottom in a list of nine categories of information. A similar result was reported by Béjoint (1981:215) where only 5% of the sample placed etymology among the three most frequently looked up categories of information. In addition, only 10.9% of Kipfer's subject (1985:62) reported looking up word history in their dictionaries.

The fact that few dictionary users look up etymology does not necessarily rule out the pedagogical role that etymological information can play in an EFL/ESP learning context. It may be that dictionary users, and probably teachers, are not fully aware of the significance of etymology in dictionaries. The question that should be asked first: does information about word origins help the language learner learn better, and if so, what type of etymological information should be considered and how? Ilson (1983:77) argues that etymology does have a pedagogical relevance and that it provides at least four types of information:

1. etyma and cognates, including borrowings and loan translations;
2. morphological analyses of lexical units in terms of their constituent structure;
3. morphological analyses of lexical units in terms of processes of word formation;
4. analyses of lexical units in terms of the cognitive procedures (e.g. metaphor, historical allusion) of their formation.

Ellegård (1978:236) criticised two of the well-known learners' monolingual English dictionaries, i.e., the OXFORD ADVANCED LEARNER'S DICTIONARY OF CURRENT ENGLISH and the LONGMAN DICTIONARY OF CONTEMPORARY ENGLISH, for having failed to "supply the learner with information that may be of considerable help in his actual learning". In Ellegård's opinion, etymological information could help the learner store the word in his/her memory through association. Another advantage is "increasing the learner's grasp of both the phonetic and the semantic structure of the word". However, Ellegård suggested that etymological information that has no 'practical aim' should be left out. However, in the area of derivations and compounds, Ellegård added, etymology is directly helpful, particularly with vocabulary of Latin and Greek origin.

Much of the medical/nursing technical vocabulary is derived from Latin and Greek. It should be noted that the ESP learner is not interested in historical accounts of word development, but in how to use his/her knowledge of word structure for remembering and making new words. Therefore, exhaustive etymological information can be counterproductive as it may demotivate the learner.

CHAPTER SEVEN
DICTIONARIES IN USE: A CRITIQUE

7.1 Introduction

Whether we approve or not, ESP learners use dictionaries in their learning and will probably continue to do so. ESP teachers should be aware of what dictionaries their students use, or do not, if they are safely to assume that their students are using the most appropriate dictionaries, i.e., the sorts of dictionaries that are capable of meeting the learner's specific communicative and lexical needs. In this chapter, I present and discuss the findings pertaining to the particular dictionaries that the students surveyed claimed to be using in their university work. I also provide the data found on the students' attitudes towards dictionaries and dictionary use and relate this to findings on the attitudes of the nursing as well as the ESP staff. Some of the students' reported strategies in obtaining lexicographical information are also presented here. However, the focus of this chapter is laid on the dictionaries themselves which are analysed and critically assessed from an ESP practitioner's point of view.

In the students' questionnaire, the subjects were presented with a list of 16 names of dictionaries of various types, sizes and orientations. They were also instructed to indicate how often they were using each of the dictionaries in the list. The dictionaries themselves represent a variety of types and sizes, including general and technical, bilingual and monolingual dictionaries. Some of them are learner-oriented and some are not; some represent British lexicography and some are American. The purpose was to provide the subjects with a variety of dictionary titles they were most likely to be using in their university studies. For lack of space in the questionnaire, it was not possible to make the list longer. Instead, two blank lines were provided at the end of the list and students were requested to mention any other dictionaries they were using and the frequency of their use. The dictionaries in the list were particularly decided upon after checking with a number of both students and staff from the Faculty of Nursing. They were asked for a subjective opinion as to what dictionaries the students surveyed were using. The pilot-test version of the students' questionnaire also gave some indications of the sorts of dictionaries that might be in current use; this in addition to my own observation of the sample first- and second-year nursing students whom I taught during the data-collection stage of the research (first semester, 1986/87).

7.2 The Image of the Dictionary

The way dictionaries are perceived by students and their teachers is likely to have bearing on the extent to which dictionaries are bought or used. In foreign language learning, students' and teachers' attitudes are crucial factors. An 'attitude' is defined by Rokeach (in Smith 1982;20) as "a relatively enduring organisation of beliefs around an object or a situation, predisposing one to respond in some preferential manner". According to Smith (ibid.: 21), there are four components to attitudes: the cognitive, the affective, the evaluative, and the behavioural. This definition of attitudes, in my view, can be related and applied to the use of dictionaries. Students and teachers develop a concept of a dictionary or dictionaries and may, for instance, think of a dictionary as merely a book of meanings of words arranged alphabetically; also they may think that all dictionaries are the same. Further, they may associate particular affective feelings with the process and experience of dictionary use; they may feel bored or frustrated, happy or excited while looking up a piece of information. The evaluative dimension occurs when students or teachers assess their own feelings about a dictionary and say, for example, "I don't like this dictionary because it is too heavy". The users then act accordingly and decide, for instance, to buy one particular dictionary, or to stop using another. Students' attitudes may also be influenced by the attitudes of others such as their teachers, parents, or fellow students.

The majority of students (89.3%) seem to appreciate the role of dictionary use in their learning as 302 students said they agreed with the statement: "It is important for nursing students to use dictionaries in their academic study", while only 30 (7.4%) disagreed and 13 (3.2%) were undecided This view is shared among the students across all levels represented in the sub-samples and may also reflect the students' dissatisfaction with their language abilities. In the follow-up interviews, on a scale of 5 points ranging from 1=not important to 5=extremely important, out of 34 students, 20 (58.8%) rated dictionary use in general as 5, 8 (23.5%) rated it as 4, whereas 5 (14.7%) gave it the rank of 3.

This attitude was confirmed through the students' reactions to the statement: "Every nursing student should have at least one dictionary". Nine students in every ten seem to share the opinion that acquisition of at least one dictionary is necessary. This view is shared among the majority of students across all levels represented, as the results show little variation between one sub-sample and another. In addition, 18 of the nursing teachers (75%) agreed with the statement above, whereas only 6 teachers (25%) reported their disagreement.

Furthermore, more than half the sample seem to be willing to consult more than one dictionary as at least 55% of the students agreed with the statement: "To depend on <u>one</u> dictionary is <u>not</u> enough"; the statement actually came from a student during the pilot-testing of the survey. There also seem to be differences between the sub-samples. For example, whereas only 41.7% of the third-year students agreed, 63.5% of the second-year agreed with the above-mentioned statement. As for the nursing staff, 87.5% (N=21) disagreed with an opposite statement, i.e., "Students should depend on one dictionary" while only 3 teachers (12.5%) agreed with it.

However, despite the fact that many students recognise the importance of owning and using dictionaries in their learning, still almost half of the sample (46.1%) seem to think of dictionaries as 'necessary evils', so to speak. Responding to the statement: "In general, dictionary using is boring", 45.4% of the students agreed while 46.1% disagreed and 8.4% could not make up their minds. But, it may be unfair to generalise this finding over all types or levels of students as the figures show wide differences between one group and another. The first-year students' responses, for example, seem to be quite different from those of the fourth-year students. Among the 24 nursing staff surveyed, it was found that 20 (83.3%) disagreed with the statement: "Using dictionaries, in general, is boring", whereas only 4 (16.6%) said they agreed.

As to dictionary typology, the students seemed to be in favour of using a medical/nursing dictionary. The students' questionnaire revealed that the majority of the sample (86.4%) agreed with the statement: "It is important for nursing students to use a medical/nursing dictionary in their study". This suggests that the majority of students face problems with technical lexis and that they need dictionaries that are capable of meeting this aspect of their needs.

It was interesting to find that almost two thirds of the students (65.6%) were against the idea that pocket dictionaries should be avoided. The students' reactions to the statement: "Nursing students should avoid using pocket dictionaries" were as follows: 15.8% agreed, 65.6% disagreed, and 18.5% were undecided. The views of the nursing staff seemed to match those of their students concerning this issue; 79.1% of the nursing teachers (N=19) disagreed with the statement: "Students should avoid using pocket dictionaries". For particular academic communicative contexts, e.g., clinical training and lectures, a small dictionary for quick reference can be useful. In fact, the nursing staff specifically ask their students to buy a pocket monolingual English dictionary, i.e., BAILLIERE'S NURSES' DICTIONARY by Kasner and Tindall, London: Baillière Tindall, 1912.

The nursing students were also found to be in favour of monolingual English dictionaries rather than bilingual ones.

The results indicated that more than half the sample (51.6%) agreed with the statement: "English-English dictionaries are more useful to nursing students than English-Arabic dictionaries", whereas only 35.3% disagreed with the same statement. The results of the Needs Analysis also confirmed that a high percentage of students (74.5%, N=302 out of 405) said they needed help regarding the use of monolingual English dictionaries. Not only did the students say they wanted help in using monolingual English dictionaries, but also they expressed their desire that the current ESP courses would include teaching materials on the use of this type of dictionary. The majority of the sample (79.7%) agreed with the statement: "It would be better if the present English language courses included exercises on using English-English dictionaries". These results probably reflect the students' dissatisfaction with their bilingual English-Arabic dictionaries and their willingness to convert into the use of monolingual English dictionaries provided that they are trained in how to use them.

The nursing teachers' questionnaire yielded further information on their attitudes to dictionaries and the teaching of reference skills. Most of them (75%, N=18) agreed that their students should be taught how to use dictionaries; in addition, 62.5% of the teachers (N=15) agreed with the statement: "Students should be encouraged to use English-English dictionaries". Moreover, the majority of these teachers (83.3%, N=20) agreed with the statement: "Students should know meanings of words/phrases in English". The nursing teachers' attitudes were in line with the students' views as far as the use of monolingual English dictionaries was concerned. In general, there seemed to be a realisation that, although bilingual dictionaries were more frequently used than monolingual ones, monolingual English dictionaries were favoured. The problem seemed to be the lack of awareness and/or proficiency in using monolingual English dictionaries rather than the attitudes towards them.

As to the nursing teachers' views on technical vs. general dictionaries, the results showed that priority was attached to technical monolingual English dictionaries as two thirds of the teachers (66.6%, N=16) said it was 'very important/extremely important' for their students to use such type of dictionaries. Coming next were English-Arabic general dictionaries; 54.1% (N=13) described using these dictionaries by their students as 'very important/extremely important'. Table 10 below summarises the results concerning this issue:

All of the interviewed ESP staff said that it was important for their students to use dictionaries in general (four of them described dictionary use as very important in the learning process). On a five-point scale ranging from 1 = not important to 5 = extremely important, the responses concerning the ESP teachers' attitudes as to dictionary typology showed that they attached most importance to the following types of dictionaries (i.e., 4 out of 6 rated each of these types as 4 or 5 on the scale).

1. large-size general English-English;
2. pocket general English-Arabic;
3. large-size technical English-Arabic;
4. pocket technical English-English;

However, Arabic-English dictionaries seemed to be unfavoured.

In conclusion, the overall tendency among the teachers, as well as the students, seems to be in favour of monolingual English dictionaries. The dictionaries needed are of both types, technical and general nature. In addition, there seems to be a case for small-size dictionaries, general and technical.

7.3 Acquisition of Lexicographical Information

I mean by lexicographical information, information on types of dictionaries, symbols and conventions employed in dictionaries, and how to use particular dictionaries; whether students try to obtain such information, and if they do, from which sources; and the roles of the ESP teachers, the subject teachers, fellow students, and the dictionary itself in providing such information.

Table 10 : Nursing teachers' attitudes as to the importance of using certain dictionary types by their students

Type of Dictionary		Attitude: not important	slightly important	very important	extremely important	I don't know	Total
English-Arabic technical	N	1	14	6	3	-	24
	%	4.1	58.3	25.0	12.5	-	100
English-Arabic general	N	4	6	8	5	1	24
	%	16.6	25.0	33.3	20.8	4.1	100
English-English technical	N	-	7	8	8	1	24
	%	-	29.1	33.3	33.3	4.1	100
English-English general	N	4	15	3	2	-	24
	%	16.6	62.5	12.5	8.3	-	100
Arabic-English technical	N	8	11	3	2	-	24
	%	33.3	45.8	12.5	8.3	-	100
Arabic-English general	N	10	6	7	1	-	24
	%	41.6	25.0	29.1	4.1	-	100

7.3.1 Asking Fellow Students about Dictionary Matters

Students refer to each other when in need of information on choice of dictionaries and/or on how to use them. Half of the sample (51.1%) said they asked each other what type of dictionary they should use. In addition, 47.9% of the sample reported that they asked each other about symbols and conventions used in dictionaries. The results, however, vary from one level of students to another. However, only 34.3% of the sample, mainly second-year students, reported asking each other on how to use particular dictionaries. The frequency of asking fellow students is generally low, though varying from one sub-sample to another. On the other hand, the results of the follow-up interviews showed that only 8 students (19.5%) said they asked other students about dictionary matters in general.

7.3.2 Asking ESP Staff about Dictionary Matters

Just above half the sample (52.6%) reported asking their ESP teachers about choice of dictionary type. As for information on dictionary symbols, almost half the sample (49.6%) said they asked the ESP teachers. For information on how to use certain dictionaries, 40.7% of the students said they asked their ESP teachers; this percentage is higher than that reported by students (34.3%) who asked other students for the same type of information. The frequency of using this strategy, i.e., asking ESP staff is still rather low.

The interviews with the 6 ESP teachers, on the other hand, showed that 5 of them said they never gave information and/or exercises on dictionaries and their use in their ESP classes as part of their teaching, and that they gave information of this sort only if a student asked. I also asked the ESP staff whether they tested their students on knowledge of, and ability in, using dictionaries; none of the teachers gave a positive answer. Furthermore, when asked whether or not they encouraged their students to use dictionaries, 5 of them said 'yes', but they said they did so 'generally'.

7.3.3 Asking Nursing Staff on Dictionary Matters

Concerning information on the type of dictionary to use, 42.4% of the sample said they asked their nursing teachers. 43.7% of the sample reported asking their nursing teachers about symbols used in dictionaries. As far as information on how to use particular dictionaries is concerned, only 26.6% of the sample indicated asking their nursing teachers. On the other hand, the follow-up interviews with the students showed that only 9.7% (N=4) of the sample reported asking their nursing teachers on dictionary matters in general.

The results of the nursing staff's questionnaire showed that all of the 24 respondents said they encouraged their students to buy certain dictionaries; 45.8% of these teachers reported doing so with a 'high' frequency, whereas 41.6% of them reported an 'average' frequency. The results also revealed that 21 teachers (87.5%) said they did advise their students on what dictionary/ies to purchase; among these 21 teachers 7 (33.3%) said they did so with a 'high' frequency, whereas 8 (38.1%) reported an 'average' frequency. As to the question whether or not they explained to their students how to use dictionaries, 17 (70.8%) said they did so; 9 (52.9%) of these 17 teachers reported a 'high' frequency, whereas 4 (23.5%) indicated an 'average' frequency.

7.3.4 Reference to Dictionary Front Matter

Familiarity with a dictionary introduction is essential and a crucial factor in successful dictionary use. I also think of a dictionary introduction as a major criterion in dictionary design and evaluation, and therefore I asked the students whether or not and if so, to what extent, they referred to such information in the dictionaries they used. The results of the students' questionnaire showed that 51.3% (N=208 out of 405) of the sample reported consulting a dictionary introduction; more than a third of them (38.4%, N=80 out of 208) indicating a 'rarely' frequency. The percentage of students among the first- and second-year students indicating a rare frequency is even over 44%. The follow-up interviews revealed that only 35% of the sample (N=14 out of 40) said they consulted the introductory information in the dictionaries they were using.

This finding can be compared to other findings reported in the literature of dictionary use. For example, Iqbal (1987:54) reported that the perusal of front matter in the dictionary among his subjects was found to be inconsistent. Hartmann (1983b:198) reported that the introductory material was rarely studied for guidance and that learner-users did not make an effort to memorise the various codes, symbols and abbreviations in their dictionaries. Similarly, Béjoint (1981:216) reported that 89% of the students in his sample answered that they had read the introductory matter less than thoroughly. In addition, Kipfer (1985:64) found that only 5.1% said they had read the preface in their dictionaries and only 6.1% said they had read the explanatory notes in the dictionaries they used.

It can be said then that a major tendency among dictionary users is their lack of interest in the dictionary front matter. This may be due to the lack of awareness of the importance of understanding the dictionary introductory matter among the users. The responsibility of creating such awareness should be shouldered by the language teachers and the syllabus/ materials designer. The lexicographer should also share the responsibility by giving more thought to the design of the dictionary front matter in terms of size, content, and format. A long introduction with irrelevant and/or sophisticated materials may put the user off. What is worse is to find no introductory material in a dictionary. The dictionary introduction should mirror the whole dictionary, its philosophy, main features, and methodology of using it. Charts that explain entries and entry arrangement, and even exercises should be considered. The use of colour is certainly an advantage in the design of a front matter. In addition, it helps the learner-user if the dictionary front matter is written in his/her mother tongue together with illustrative examples on how to make most use of the dictionary.

7.3.5 Browsing through Dictionary Pages

Browsing through dictionary pages without the specific purpose of looking up a particular piece of information is a habit that plays a role in familiarising the user with the dictionary. More than three quarters of the students (78.5%, N=318 out of 405) in this study reported using this strategy, above a third of them (35.5%, N=113 out of 318) said they did so with a 'high' frequency. The percentages of students indicating use of this strategy are particularly high among the first- and second-year levels. The results of the follow-up interviews also showed that 72.5% of the sample (N=29 out of 40) said they browsed through the dictionaries they used. In her study, El-Badry (in progress) found that 67.5% of her sample said they liked browsing in dictionaries. Béjoint (1981:216) reported that 55% of his sample said they sometimes browsed through their dictionaries without looking up anything in particular.

Both dictionary designers and course planners should capitalise on this habit among users. The dictionary should include materials and employ methods that are likely to create and/or enhance the user's curiosity and interest in the dictionary. Usage notes and the inclusion of drawings, diagrams and pictures may add life to the process of dictionary use. Also, language teachers should encourage their students to browse and read through their dictionaries since dictionaries are educational tools in themselves and have a role in enriching the students' learning experience.

7.4 Students' Use of Bilingual Dictionaries

7.4.1 AL-MAWRID

Of the 16 dictionaries listed in the students' questionnaire, a general bilingual English-Arabic large-size dictionary was found to be the most popular. This dictionary is named AL-MAWRID and compiled by the Lebanese translator Munir Ba'albaki. The figures showed that 79.7% (N=323 out of 405) claimed to be using this dictionary. Among these users 58.5% (N=189 out of 323) reported a 'high' frequency of use, in addition to the 25.6% (N=83 out of 323) who reported an 'average' frequency. The highest percentage of users is noticed to be among the fourth-year students (91.1%, N=93 out of 102), but the first-year students seem to be the students who use it most frequently (66.6%, N=60 out of 90). The follow-up interviews also showed that 73.1% of the sample (N=30 out of 41) said they were using AL-MAWRID.

To explore the students' attitudes towards this particular dictionary, I asked them to react to the following statement: "AL-MAWRID is the most useful dictionary for nursing students". Almost two thirds of the students (64.4%) agreed with the statement whereas only 12.1% disagreed, while 23.4% could not make up their minds. This result confirms the popularity of AL-MAWRID among the students. However, the finding that 12.1% of the students disagreed with the above-mentioned statement, and that 23.4% of them were undecided means that more than a third of the sample (35.4%) had reservations about AL-MAWRID. The same can be said about the nursing teachers' responses concerning AL-MAWRID. The nursing staff were asked to react to a similar statement: "AL-MAWRID (large-size) is the dictionary that the nursing students should use". Just over half of the nursing teachers (54.1%, N=13) expressed their agreement, whereas a third (33.3%, N=8) disagreed, while 12.5% (N=3) were undecided. Furthermore, 16 nursing teachers (66.6%) gave the title AL-MAWRID as the English-Arabic dictionary they liked to recommend for the students to use.

As for the ESP staff, all of them said they suspected that their students were using AL-MAWRID as the main dictionary. When asked what dictionaries they would like to recommend for their students, 4 of the ESP teachers mentioned AL-MAWRID, whereas 2 of them said they found it difficult to recommend any particular dictionary as they themselves lacked knowledge of dictionaries. AL-MAWRID was similarly found to be the dictionary most relied upon by the 13 first-year students who participated in the Dictionary Using Diaries experiment. El-Badry (in progress) and Al-Besbasi (in progress) similarly found that AL-MAWRID was the most popular dictionary among the users they surveyed.

To highlight the underlying features and design principles of this dictionary, I examined its 1986 edition. The following is a critical review and discussion of these features. The first edition of AL-MAWRID was published in 1967 in the Lebanon by Dar EL-ILm LiL-Malayen. The compiler, Munir Ba'albaki, wrote in the preface (all in Arabic) that he had worked as an English-Arabic translator for more than thirty years. According to him, seven years were spent in producing his dictionary. The compiler thus wrote his dictionary from a translator's perspective rather than from that of a foreign language teacher. As a translator, Ba'albaki felt dissatisfied with the bilingual dictionaries available at the time. He stated in his preface that the dictionaries he used were in varying degrees far from being able to meet the needs of the translator. Another motivation to produce AL-MAWRID was the increasing expansion of modern science to an extent that made the then available English-Arabic dictionaries, as the compiler put it, incapable of meeting the requirements of the age.

The dictionary was based on American and British monolingual dictionaries, in addition to a set of bilingual general and technical ones. The monolingual English sources included the OXFORD ADVANCED LEARNER'S DICTIONARY OF CURRENT ENGLISH (London, 1963), but no mention was given of any monolingual technical dictionaries. No indication was given as to how the compiler benefited from the learner's dictionary, in particular. The bilingual source dictionaries included science and technology dictionaries and technical glossaries produced by the Arabic Academy in Cairo. These covered subjects such as botany, astronomy, agriculture, forestry, education, psychology, military science, and medicine (the medical source dictionaries were SHARAF MEDICAL DICTIONARY (Cairo, 1926), and HITTI'S MEDICAL DICTIONARY (Beirut, 1967), both large-size English-Arabic). Despite the fact that AL-MAWRID was based on a wide range of dictionaries, there was no mention of any reference to any research in related fields such as contrastive linguistics, lexicography, or foreign language teaching. Hence, the main input to AL-MAWRID came from the source dictionaries and the compiler's own translation experience.

Concerning the usership, the compiler did not state clearly who the specific users might be, although it would be possible to guess that translators were meant to be the main target. There was, however, some reference to the frustration and the disappointment of the 'educated Arab' whenever s/he refers to English-Arabic or English-English dictionaries. The only direct reference to the audience appeared in the last paragraph of the preface where the compiler hoped that his dictionary would gain the satisfaction of researchers, teachers and the generally educated public.

The dictionary is claimed to be comprehensive and capable of saving the 'educated Arab' from continuous reference to other dictionaries. According to the compiler, AL-MAWRID has no less than 100,000 entries "representing most of the English language and the terminology of modern science and arts of the human civilisation". It should be noted that derivatives and compounds are also listed as independent entries.

According to Ba'albaki, AL-MAWRID was based on seven design principles. These are translated and presented here in paraphrased form.

1. Entries are presented first in the different forms of spelling, e.g., <u>equivoque</u> or <u>equivoke</u>, followed by pronunciation which is indicated by special symbols, based on THE AMERICAN COLLEGE DICTIONARY (n.d.). Then follow the part of speech and etymology, especially when

the entry is of Arabic origin. When necessary, this is followed by one or more forms of the entry's plural with an indication of the pronunciation of the plural in most cases.

2. The meanings/senses are presented according to the arrangement of the parts of speech mentioned at the beginning of the entry. For instance, if the parts of speech are entered as noun, verb, adjective, the meanings of the noun are given first, then the meanings of the verb and so on. The symbol § is used to indicate a shift from a particular part of speech into another whereas the symbol x is used to signal the shift from a transitive to an intransitive form of verb or vice versa.
3. Within the entries, a historical approach is adopted in the listing of meanings. This is, the compiler argues, to make it possible for the user to trace the historical development of the word. Relying on WEBSTER and OXFORD dictionaries, i.e. WEBSTER'S THIRD NEW INTERNATIONAL DICTIONARY OF THE ENGLISH LANGUAGE (London, 1961), WEBSTER'S SEVENTH NEW COLLEGIATE DICTIONARY (Springfield, MA 1965) and THE SHORTER OXFORD ENGLISH DICTIONARY (Oxford, 1964) - obsolete words as well as American and British colloquial words are labelled.
4. The senses of the entry are classified into sets or families, hence synonyms are presented in one place. Every set is given a serial number, and the sub-senses are distinguished alphabetically in Arabic.
5. Before entering a sense or a set of senses, the source dictionaries are consulted to check how each treated an entry. This aimed at getting a picture of the whole meaning and its shades so that the Arabic equivalent would be made as close to the English meaning as possible. Ba'albaki argued that a one-word Arabic equivalent might be sometimes misleading.
6. Arabic equivalents, however precise, would not be sufficient to clarify the intended meaning, in the compiler's point of view. Thus, entries in AL-MAWRID

include examples quoted from the source dictionaries and indicated in brackets. This is aimed at making the dictionary "a useful tool for those who want to write in English, in addition to its role as a reference for those who seek the meaning of a word or a phrase".

7. The terms of modern science and contemporary civilisation are entered in AL-MAWRID. Techniques in treating the terminology included arabicisation (finding or creating an Arabic equivalent), transliteration and derivation. It is claimed that precision and simplicity were taken into account and that in most cases a brief explanation which gives the user an idea, however short or simple, about the meaning is given. The branch of science or art to which a term belongs is indicated through Arabic initials. The advantage of this, it is argued, is that it provides an encyclopedic touch in treating terminology which is often missing in the English-Arabic dictionaries that preceded AL-MAWRID. (Ba'albaki is, in particular, critical of English-Arabic medical dictionaries because their compilers were content with mentioning the equivalent term without any attempt to provide an explanation and/or a definition.)

Prefixes and suffixes are said to be treated with special attention because they are seen as "indispensable keys to the understanding of thousands of compound words", and therefore they are entered as independent entries. Their meanings are given and one-word examples are usually provided. Moreover, the dictionary includes idioms in addition to colloquial words said to be common in the works of English and American contemporary writers. Use is made of cross-referencing to save space and the offset printing technique is claimed to have been used for the first time in English-Arabic lexicography.

The dictionary preface is followed by 'general guidelines' where the user is shown an example entry and how to read it. Some symbols and dictionary conventions are also explained in these 'guidelines'. A pronunciation key is provided and the key to the pronunciation of vowel sounds is reproduced at the bottom of every page in the dictionary. The introductory matter also includes a list of the Arabic and English abbreviations used in the dictionary.

AL-MAWRID is republished annually and claimed to be updated. New words entering English, particularly in science and technology, are added, e.g., cytopathology and cytoscopy which first appeared in the dictionary in 1986. Stress is indicated through rather unusual symbols; a thick comma signals the main stress whereas a thin comma indicates the minor stress. Grammatical information is shown through abbreviated labels, e.g., (adj. for adjective). These include parts of speech, whether a verb is transitive or intransitive, whether a noun is plural or singular, and whether an article is definite or indefinite. Abbreviations and derivatives are treated as independent entries, but phrasal verbs are listed under the main verb entries.

The use of colour is another feature of the dictionary. Coloured pictures and drawings of various sizes appear on almost every page. In addition, the guide words at the top right and top left of every page are printed in red, as is the pronunciation key at the bottom of each page. Almost in the middle of the dictionary there is a set of eight plastic transparent and coloured plates showing the anatomy of the human body. The body parts are numbered and on either side of the plate the numbers and the parts they indicate are listed. All names of the body parts are given Arabic equivalents and an English index (showing the numbers of the plates, the numbers of the parts and their names in English) is also provided. This feature is of particular relevance to nursing and medical students. However, many of the Arabic equivalents are rather unintelligible. This is probably due to the fact that these terms are either invented or derived from classical standard Arabic, a variety which is not in current use, especially in the spoken medium. In addition, the English index lacks any guidance on pronunciation.

Another encyclopedic aspect is the provision of eight glossy coloured plates of animals, birds, fish, flowers, fruits, gemstones and vegetables. These are placed towards the middle of the dictionary. The names of the animals, birds, etc. are given in English and Arabic, but no help with pronunciation is supplied, possibly because it is assumed that the user can refer to the main entries in the dictionary. At the end of AL-MAWRID, there is quite a lot of information in appendix form. There is a 28-page list of customary abbreviations used in modern writing; for these the Arabic translation is given and the user is sometimes referred to the page on which the abbreviation appears as an entry. In addition, a seven-page list of irregular verbs in English is given as an appendix. (Irregular past tense and past participular forms of verbs are treated as separate entries in the main body of the dictionary.) The user is also provided with a rather lengthy appendix of a 95-page collection of English proverbs and their Arabic equivalents. A further appendix of English words of Arabic origin (12 pages) is included. The dictionary ends with a 93-page biographical appendix with coloured pictures, pronunciations and a brief synopsis of each personality involved. Thus, the user of AL-MAWRID is supplied with an extra 235 pages of appendixed information of an encyclopedic nature.

Having reviewed, in some detail, the main features of AL-MAWRID, being considered the most popular dictionary among the sample, it is now possible for me to make some observations from an ESP point of view.

1. The dictionary was first published in 1967. Remembering that it took the compiler seven years to complete it, we can conclude that it started almost at a time when ESP as a discipline was just evolving. The fact that the dictionary is republished annually does not necessarily imply that it is up to date, despite the fact that some new technical terms, new and different coloured pictures at the beginning of every letter and new appendices are added occasionally. I have compared the 11th edition (1977) with the 20th edition (1986) to discover the extent of the difference, but the content and format of the preface and the introductory matter are still

exactly the same. The 1986 has different coloured pictures introducing each letter of the alphabet; the 1977 edition has the same animal, bird, fish etc. plates but these plates are located in different places in the dictionary whereas they are grouped together in the 1986 edition. The 1977 edition has no more than two appendices at the end (the abbreviations and the irregular verbs). It also lacks the human body plates which are entered in the 1986 copy. Almost all other design features remain the same. From this it can be concluded that AL-MAWRID is relatively old and draws upon no principles taken from ESP teaching.

2. AL-MAWRID was compiled from a translator's perspective and there appears no evidence that the needs of EFL/ESP learners were in any way taken into account.
3. The compiler has relied upon his own experience in translation and on a set of dictionaries which included only one learner's dictionary. In addition, no reference is made to any monolingual technical dictionary.
4. The terminological input in the dictionary was based on English-Arabic technical dictionaries and glossaries by the Cairo Arabic Academy. In many cases, the compiler has taken the initiative to 'terminologise' in addition to relying on arabicisation rules. It can be argued that the technical dictionaries, on which AL-MAWRID was based, were glossaries themselves, as we shall see later in this chapter. A distinction must be made between a dictionary aimed at learners of English as a foreign language and a glossary whose aim is to contribute to the arabicisation of science and technology.
5. The usership of AL-MAWRID is not clearly specified. It is rather aimed at a wide range of users: translators, researchers, teachers and the educated Arab in general. The term 'educated Arab' stressed in the preface of the dictionary covers every possible user. From an ESP

point of view, the dimension of user-specificity is missing in the dictionary.

6. AL-MAWRID is supposed to be a general, all-purpose dictionary, and meant for both receptive as well as productive users of the language. The content of general English is quite wide, "to represent every art of human civilisation". Much of this content is irrelevant to an ESP learner and therefore it can be said that the dimension of being purpose-specific is also missing in the dictionary.
7. Even with its terminological content, the dictionary tries to cover all branches of modern science. Though some medical content is provided, this is only a small ratio of the science content. The ESP learner is, by definition, concerned with learning the English used in one specific area of human knowledge, e.g., medicine or nursing; thus botany, astronomy, forestry, agriculture, military science etc. become redundant from the nursing student's point of view.
8. The historical approach adopted in arranging the meaning of entries is in no way relevant to the interests of ESP learners. Few of the ESP nursing students in this study would be interested in "tracing the development of the word to get a picture of how human civilisation was developing through history". Rather, they would concern themselves with how commonly used words might be and with understanding and using such words appropriately in academic contexts.
9. At the macro-structure level, i.e., the overall organisation of dictionary materials, a large percentage of headwords is unlikely to be of direct use to a particular group of ESP students. They have to scan a whole page and skim through many entries in order to get to their target words. Furthermore, only a few of the 235-page appendices would be of specific help to an ESP learner; the appendix of the irregular verbs in English and the appendix of human body plates are exceptions.

In general, the ESP learner has to handle and search through a thick and heavy book of 1,353 pages whenever s/he wants to look up a particular word.

10. At the micro-structure level, i.e., the organisation of information in an entry, it can also be argued that much of the content provided would be redundant as far as the ESP learner is concerned. For example, while very little information is given on technical terms, there is much irrelevant, and probably distracting, information on a whole host of general English vocabulary of specific meaning which may be described as sub- or semi-technical. Such words are usually entered in all-purpose monolingual and bilingual dictionaries because of their general nature, therefore all possible meanings are given. The ESP learner would be particularly concerned with the meaning that pertains to his/her subject area and the rest of the information becomes irrelevant. To take an example, a nursing students of ESP may want to look up one of the following words: admit (admission), airways, apply, beat, cell, charge, discharge, congestion, contract, depression, labour, culture, dispatch, theatre, cuff, discipline, disorder, inspiration, shock, stress, vessel. All these words are entries in AL-MAWRID, but to get to the medical/nursing meaning, a nursing students has to read through a long entry and work out for him/herself the specific meaning. Such words are not labelled in any way that may help the learner to locate their specific meaning.

Most of the students in this survey said they were using AL-MAWRID with a relatively 'high' frequency. However, the question remains as to whether AL-MAWRID is the 'right' dictionary that the students in this sample should be using. I have argued that AL-MAWRID is in no way, and has never been, intended for an ESP learner target. The students are using it and may continue to do so, not because AL-MAWRID is the most appropriate dictionary for them, but rather because it can be considered the best available general bilingual dictionary on the market. What makes the students rely on it may be explained by the following points:

1. AL-MAWRID is generally comprehensive and its coverage of medical terminology is an advantage. The nursing students at this stage are learning both their subject matter as well as the English language, and in many cases, they are proficient in neither of these, unlike doctors or nursing practitioners who have already acquired at least the subject matter. This means that the students need access to both general, but relevant English words and medical/nursing terminology. AL-MAWRID does provide help, though limited, in both areas. In comparison to other bilingual general or technical dictionaries or glossaries, it treats the medical content in a more helpful manner. For example, it labels the medical term and the branch of medicine it may belong to, such as surgery or pathology. It also provides a brief indication of the meaning of the term.
2. Another advantage that AL-MAWRID has is its wide availability in the Arab World; in Jordan it is available in almost every library or bookshop and certainly in the libraries of the University of Jordan and the Faculty of Nursing.
3. It is also comparatively cheap for its size and format. The 1987 edition was priced at 3.5 JD (Jordanian Dinars), an equivalent of £6 at the time when my fieldwork was being carried out.
4. The provision of illustrative examples within entries is a positive characteristic of AL-MAWRID. Many existing bilingual dictionaries lack this important feature.
5. AL-MAWRID provides information on grammar, though limited, usage and pronunciation (and the reproduction of the pronunciation key at the bottom of every page adds to its credit.
6. The use of colour is a further credit to AL-MAWRID, together with the use of drawings, plates and pictures.

7. The human body plates may appeal to the nursing student and also the appendix on irregular verbs in English.
8. AL-MAWRID is usually the first dictionary to come to the minds of the members of ESP and nursing staff when asked to recommend a particular dictionary. Students normally listen to their teachers and take their advice, however uninformed that may be.

7.4.2 HITTI'S MEDICAL DICTIONARY

The second most popular dictionary among the students was also found to be a bilingual one, but technical this time. This is HITTI'S MEDICAL DICTIONARY (English-Arabic, large-size) by Usuf Hitti, first published in 1967 by Librairie Du Liban. 60.2% (N=244 out of 405) of the students said they were using it, and 56.1% (N=137 out of 244) of these students indicated a 'high' frequency of use. To find out what HITTI'S offers to ESP nursing students, I inspected the "fourth revised edition of 1982, new impression 1984". What follows is a critical assessment of its main features. The compiler Usuf Hitti, a Lebanese medical doctor, mentioned in the preface to the first edition that it took him seven years to complete his dictionary. He saw his lexicographical work as a contribution in the service of Arabic. He also hoped that his dictionary would serve as a reference tool for Arab doctors, those who studied medical sciences in Arab and other institutions through the medium of English, or those studying medical sciences in the University of Damascus through the medium of Arabic. Clearly Hitti did not differentiate between the needs of a medical student and the needs of a medical practitioner. There was also no distinction between the needs of users who were studying, or have studied, medicine in English and those who were studying it, or have studied it, in Arabic. In addition, the compiler made no mention of nurses, or student nurses, as targets of his dictionary. According to Hitti:

> "If a doctor or a researcher in a branch of medical science wants to do research or make a study; or write an essay on medical topics in Arabic, he will find in this dictionary, the English terms arranged alphabetically, and their equivalent meanings in Arabic"
> (preface, translated from Arabic)

Thus, we have a dictionary that sets its major goal as a contribution to the Arabic language, and therefore it can be considered as an exercise in arabicisation rather than pedagogical lexicography. It is intended for doctors and medical researchers who may want to write in Arabic; and/or students of medicine who are studying through the medium of Arabic in a particular location, i.e., Syria. The purpose of this dictionary remains to promote proficiency in Arabic rather than in English. In our case, we are concerned with undergraduate nursing students who are learning their subjects through the medium of English. Therefore, it is difficult to accept the assumption that simply because the dictionary is about medicine it would be the 'right' reference tool for this specific group of users we are concerned with in this study.

The English content in HITTI'S MEDICAL DICTIONARY was based on two monolingual English technical dictionaries: DORLAND'S ILLUSTRATED MEDICAL DICTIONARY, editions 23 (1960) and 24 (1965), published in London & Philadelphia, and the FABER MEDICAL DICTIONARY (London, 1953). On the other hand, in 'arabicising and translating' the English medical terminology, Hitti referred to old medical books (written in Arabic) by Arab and non-Arab writers, and to books by Arab doctors who taught medicine through Arabic in Syria and Egypt. Among his sources were also the publications of the Arabic Academies of Damascus, Cairo and Baghdad. This was in addition to a large-size English-Arabic dictionary: DICTIONARY OF PHYSICAL AND MEDICAL SCIENCES, by M. Sharaf (Cairo, 1928). When it was not possible to find an existing Arabic equivalent, the compiler resorted to the rules of arabicisation prescribed by Arab linguists and educationalists, and methods such as compounding, derivation and transliteration.

In his preface to the second edition (1972), Hitti claimed that his dictionary had been well received and therefore had to be republished. The preface, again, confirmed that the main interest of the compiler was arabicisation more than anything else. Enriching the Arabic language with newly terminologised words and coining from other languages was emphasised. According to Hitti, because of lack of standardisation in medical Arabic terminology, various different Arabic equivalents were given. The 'researcher' was once again mentioned as the target user of the dictionary.

The preface to the fourth edition (1979) gave further information on the philosophy of this dictionary. It was claimed that the dictionary had been revised and more terms were added. The work of Arabic Academies was acknowledged for their contribution to modern medical Arabic. In this edition (1979), an appendix of combining forms (of Greek and Latin roots) in medical terminology was provided. Another appendix of 'abbreviations frequently met in the medical literature' was also introduced in this edition.

Apart from the short prefaces (in Arabic) described above, a portrait, probably of the compiler, a dedication and a table of contents, the dictionary does not include any introductory materials. There is no user's guide of any sort and there are no explanatory charts or phonetic keys. The dictionary is 758 pages (A4 size) long. These are followed by 31 coloured plates of the human body (reproduced from STEDMAN'S MEDICAL DICTIONARY, published by Williams & Wilkins - Baltimore, n.d.). These plates are introduced by an index of English and Arabic equivalents. The plates are glossy but not transparent like the ones in AL-MAWRID. The various parts of the human body are indicated by arrows rather than numbers. Following the plates, there is an 'English-Arabic appendix of the terms designated on the anatomical plates'. A 23-page appendix of 'combining forms in medical terminolgy' of Greek and Latin roots, English and Arabic equivalents and one-word English examples are provided. There is also a nine-page appendix of medical abbreviations but no Arabic equivalents are provided for these abbreviations. A supplement of a 2-page list of names of leading Arab physicians with indications of their principal contributions is inserted.

The purpose is to pay tribute to these medical scientists and to save the original forms of the spelling and pronunciation of their names from distortion. Only the name is given in Arabic but other information in English. The rest of the dictionary is a 106-page Arabic-English glossary. The arabicised terms are excluded from the glossary, though mentioned in the main English-Arabic body of the dictionary. This glossary is arranged according to the Arabic alphabetical system but no guide words appear at the top of the glossary pages. The compiler states that the aim of this glossary is "to bring back to life many useful medical Arabic terms of which the Arabic language heritage is full" (preface, translated from Arabic).

Whereas the dictionary provides guidance on pronunciation for the Arabic equivalents throughout, the English terms do not receive any pronunciation treatment whatsoever. This confirms the compiler's emphasis on the Arabic side of the dictionary rather than the English. Thus, the phonetic element as far as the English version is concerned, is totally missing in this dictionary. The other striking shortcoming in this dictionary, from an ESP point of view, is that the terms entered are neither defined nor explained, either in Arabic or in English. What is provided is basically a straightforward English-Arabic glossary. On the Arabic equivalent side a multiple of synonym equivalents appears due to lack of standardisation in Arabic scientific terminology.

It is doubtful whether the majority of the Arabic equivalents given in HITTI'S would be intelligible to undergraduate nursing or medical students in Arab universities, especially where the medium of instruction is mainly English. In pragmatic terms, it is not difficult to argue that the body of the Arabic equivalents in the dictionary constitute a foreign language in themselves. The ESP learner, when looking up a medical term, is interested in understanding the concept the term designates rather than in what it is called in Standard/Classical Arabic. The dictionary assumes that the user is already familiar with the Arabic terminology and that such Arabic equivalents are in current use. Neither assumption can be accepted from an ESP learning perspective. Taking, for example, the case of a first-year nursing student user looking up the term tabescence, it is very unlikely that s/he could work out what قحل (qahal), دنف (danaf), without reference to a monolingual Arabic medical dictionary, a variety which is rarely available anyway.

Moreover, HITTI'S does not offer its users any illustrative examples to clarify meanings of the terms. Added to this, no grammatical information of any sort is provided. What is also striking is the great amount of space wasted in this dictionary. There are big white gaps on almost every page, which could have been efficiently exploited to indicate useful information. Furthermore, apart from the appended plates, the dictionary does not include any drawings, pictures, and/or diagrams that may help the user. Medical and nursing terminology includes quite a few terms that designate names of instruments or processes. For example, terms such as tachometer, stethoscope could have been better shown with a drawing or a picture. What is also missing in the dictionary is the use of colour, with the exception of some of the anatomical plates.

From an ESP point of view, HITTI'S leaves much to be desired and cannot be considered as the 'right' dictionary to be used by the students surveyed in this study. What is also interesting is that even from the points of view of the nursing staff this dictionary does not seem to be highly valued as I found that none mentioned it as one they liked to recommend for their students to use. HITTI'S is not a learner-oriented dictionary and is based on no research data. It is little more than a technical glossary of medical terms matched with, either invented, or coined from Classical/ Standard Arabic, terms whose intelligibility is uncertain from an undergraduate learner's point of view and whose currency is not well established yet.

Nonetheless, the nursing students do use this dictionary, in addition to AL-MAWRID; not necessarily because they are the best dictionaries for the students' needs but basically because a better bilingual alternative is not available to them. The fact that the students said they referred to both dictionaries suggests that these learners need access to both general English as well as medical/nursing English vocabulary. AL-MAWRID provides help in both areas but it is not enough, which partly explains the resort to HITTI'S.

7.4.3 Other Bilingual Dictionaries

The students' questionnaire yielded information, Table 11, on other less popular bilingual general and technical dictionaries. It is interesting to find out that almost half of the sample (49.3%, N=200 out of 405) said they were using a general-purpose pocket dictionary: AL-MAWRID AL-QAREEB by Munir Ba'albaki, first published in 1968, Beirut, Lebanon, (a pocket version of AL-MAWRID reviewed earlier in this chapter). Nearly a third of these students (31.5%, N=63 out of 200) reported a 'high' frequency of using it, and with the exception of the first-year students, no less than 50% of the students in each year said they were using it with varying frequencies. In the preface to the 1980 edition of this dictionary, Ba'albaki mentions that this pocket version was introduced after he found out that the large AL-MAWRID was welcomed by 'the scientists and the generally educated Arab'. AL-MAWRID AL-QAREEB is intended for the 'young generation' in the eastern and western parts of the Arab World. 'Young generation' is interpreted in the preface as 'students of the English language'. There is no indication, however, of the level, or range of levels, such students are expected to be at. The dictionary covers a quarter of the large AL-MAWRID and the selection of entries is said to have been made in the light of what English and American concise dictionaries specifically prepared for the 'young generation'. There is a note by the compiler which refers the user to the large version in case of not finding an entry. The introductory material includes a list of abbreviations used and brief guidance as to the symbols employed.

Table 11 : Students' use of bilingual dictionaries

Name of Dictionary	Description		Used by	Frequency		
				low	average	high
AL-MAWRID	English-Arabic, large general purpose	N= %=	323 79.7	51 15.8	83 25.7	189 58.5
HITTI'S MEDICAL DICTIONARY	English-Arabic, large technical	N= %=	244 60.2	38 15.5	69 28.2	137 56.1
AL-MAWRID AL-QAREEB	English-Arabic, pocket general purpose	N= %=	200 49.3	74 37.0	63 31.5	63 31.5
AL-MAWRID AL-WASEET	English-Arabic medium, general purpose	N= %=	146 36.0	67 45.9	51 34.9	28 19.1
OXFORD ENGLISH-ARABIC DICTIONARY OF CURRENT USAGE	English-Arabic, large general purpose	N= %=	114 28.1	35 30.7	38 33.3	41 35.9
Elias' MODERN DICTIONARY	English-Arabic, large general purpose	N= %=	107 26.4	39 36.4	37 34.5	31 28.9
Elias' POCKET DICTIONARY	English-Arabic, pocket general purpose	N= %=	101 24.9	50 49.5	28 27.7	23 22.7
DICTIONARY OF SCIENTIFIC TECHNICAL & ENGINEERING TERMS	English-Arabic large, technical	N= %=	75 18.5	29 38.6	31 41.3	15 20.0
Elias' MODERN DICTIONARY	Arabic-English, large general purpose	N= %=	75 18.5	36 48.0	24 32.0	15 20.0
POCKET MEDICAL DICTIONARY	English-Arabic pocket, technical	N= %=	68 16.8	29 42.6	20 29.4	19 27.9

No help with pronunciation is provided but parts of speech are labelled throughout and plural forms are indicated. Coloured and black-and-white pictures and drawings are inserted here and there but no appendices are included in this version. As far as meaning is concerned, only the most common and currently used entries were selected for this dictionary, the compiler states. He offers, however, no explanation as to what criteria he used in determining the most frequently used words. What may account for the students' use of this dictionary is that in general they are so accustomed to bilingual dictionaries throughout their school education. They also need a quick-reference dictionary which saves them the trouble of consulting the large AL-MAWRID. Yet, I do not think that AL-MAWRID AL-QAREEB is a proper choice for their needs as it lacks pronunciation guidance and illustrative examples, in addition to the great amount of redundancy that it contains, being a general-purpose dictionary.

AL-MAWRID AL-WASEET, first published in 1971, Beirut, Lebanon, another version of AL-MAWRID, by Munir Ba'albaki was claimed to be used by 36% (N=146 out of 405) of the sample; less than a quarter of them (19.1%, N=28 out of 146) reported a 'high' frequency of using it. This dictionary is very much the same as, and based on, the large-size AL-MAWRID by the same compiler, except that it has half the number of entries. In the preface to the 1971 edition, as well as on the cover, it is mentioned that the dictionary is meant for school pupils at the intermediate and secondary stages of education. But, in the 1977 edition of the same dictionary, there is no mention of intermediate and secondary school pupils and instead the preface specifies that the dictionary is intended for the 'generally educated Arab'. The dictionary has no appendixed information.

Less than a third of the sample (28%, N=114 out of 405) reported using the OXFORD ENGLISH-ARABIC DICTIONARY OF CURRENT USAGE, edited by N.S. Doniach, first published in 1971, Oxford. Both the foreword and the preface of this dictionary are in English. The dictionary seems to have been compiled as a team project where both native speakers of Arabic and of English shared the effort, but there is no indication as to whether any of the lexicographers or the editor was a teacher of English as a foreign language. The usership of the dictionary is stated in the preface as 'those whose mother-tongue is English and who are learning Arabic, and ... those whose mother-tongue is Arabic and who are learning English'. This is certainly an ambitious endeavour but it is doubtful whether the needs of both types of users can be met in one dictionary. In addition, there is no mention of how the needs of such usership were determined and whether or not the dictionary was based on specific research input.

One advantage of this dictionary is that it uses illustrative phrases or abbreviated indications in brackets when a particular word has a synonymous meaning. The grammatical function of words is also indicated by abbreviating the part of speech, and idiomatic usage is also shown. Compound verbs such as take away are treated as subsidiary entries under the main verb, and common prefixes have no separate entries. Prefixes which are used exclusively in scientific and technical language are included, but only those whose use is "sufficiently wide to warrant their inclusion" as entries in their own right. The Arabic entries are said to be authentic and up-to-date and pronunciation is provided for them. However, the pronunciation of the English words is not indicated. There are no appendices in the dictionary, and colour, drawings and/or pictures are also missing. Stylistic information, e.g., 'derogatory', 'archaic', 'figurative' is indicated and subject matter such as physiology, psychology, medicine, anatomy is labelled through abbreviations.

Other less used dictionaries also included Elias' MODERN DICTIONARY (English-Arabic, large size) by Elias and Elias, first published in 1913 by Elias Modern Publishing House, Cairo and distributed by Dar al-Jil, Beirut. 26.4% (N=107 out of 405) of the sample reported using it, among them 28.9% (N=31 out of 107) indicated using it with a 'high' frequency. In his preface to AL-MAWRID, Ba'albaki refers to Elias' MODERN DICTIONARY and argues that, together with other English-Arabic dictionaries available at the time, the MODERN DICTIONARY is inadequate and incapable of meeting the translator's needs. I examined the 1986 edition and the following is a review of its features.

There are two prefaces by Elias, both undated, one in English and another in Arabic. This suggests that it is intended for two different types of users simultaneously. Elias writes in the English version of his preface that his dictionary "has been compiled chiefly with a view to the needs of the Arabic-speaking students in their study of English" but adds that it is also aimed at "English-speaking students of Arabic, who have attained a fair degree of proficiency in that language". Thus, it is similar in orientation to the OXFORD ENGLISH-ARABIC DICTIONARY by N.S. Doniach mentioned earlier. In one respect, Elias' MODERN DICTIONARY can be credited for being a learner-needs-oriented reference book. However, it is unclear how the compiler found such needs of both types of users.

The approach adopted by Elias is the use of 'modern Arabic' as opposed to 'classical Arabic' which is said to be, though rich and generous, "poetic and chivalrous rather than scientific and practical". 'Modern Arabic' is claimed to be "popular in Egypt and all other Arab countries and updated by best authorities of recent writers" but the compiler provides no evidence to substantiate his claim. Scientific and technical words are entered in the dictionary and clearly labelled by bracketing the field of science to which they belong.

In the Arabic version of the preface, Elias criticises English-Arabic dictionaries that preceded his dictionary on the grounds that they "were originally designed for English-speaking students of Arabic as their authors entered information from which the Arab student will not benefit" (translated from Arabic). I think, from an ESP perspective, the same criticism can be directed to Elias' 'modern' dictionary as one would find information in it that is irrelevant to Arab learners of English, let alone the specific needs of ESP learners. For example, while the Arabic equivalents receive pronunciation treatment, the dictionary provides no help whatsoever as far as the pronunciation of English words is concerned. In addition, despite the fact that scientific and technical words are entered and labelled, the Arabic equivalents, like in HITTI'S, are in many cases transliterations drawn from the glossaries of the Arabic Academy in Cairo. Take as an example the term affenspaete which has the equivalent (طب وتشريح) [الشق القردي]; all the Arab user of the dictionary would be able to know is that it is a word to do with medicine, as the brackets indicate.

In treating meaning, Elias says that 'real' meaning is given first then the figurative. Full stops separate synonyms, and the shift from one meaning to another, or from one verb form to another is signalled through special symbols. No attempt is made to indicate grammatical information, even parts of speech, but the compiler does recognise the role of English prepositions and therefore he italicises what he thinks necessary of such prepositions and enters them in front of some words in the dictionary. For example the verb entry 'consist' is indicated as follows consist (of). This is a positive feature from the ESP learner's point of view; however, it seems that the treatment of prepositions is based on ad hoc judgements rather than on informed research findings.

There are black-and-white pictures and drawings throughout the dictionary and these are also grouped thematically and entered at the end as a 'pictorial supplement'. Though of general interest to the general user, much of the content in this supplement is irrelevant to the medical/nursing ESP learner, with the exceptions of the optical and surgical instruments, and parts of the human body.

Elias' POCKET DICTIONARY (English-Arabic, a pocket version of Elias' MODERN DICTIONARY, first published in 1973, Cairo, Egypt) was reported by nearly a quarter of the sample (24.9%, N=101 out of 405), though half of these students (49.5%, N=50) reported a 'low' frequency of using it. The compiler, in the preface of the 1986 edition, states that this dictionary is offered to students of English as a replacement for pocket monolingual English dictionaries which do not provide the meaning s/he seeks in his/her own language. Another motivation for compiling this dictionary is that, according to Elias, many elementary school pupils had to own the large-size version although they were not in need for many of its content in their early stages of learning English. This implies that Elias recognises the principle that dictionaries should be needs-based and that different types of dictionaries are needed for different levels of learners. He also realises that, from the school learner's point of view, much of the large-size version becomes redundant.

The dictionary is targeted at both English-speaking learners of Arabic as well as Arabic-speaking learners of English. Thus, pronunciation of Arabic words is indicated, but it is ignored in the case of English words. Almost all other aspects of the large version apply to this pocket one, with the exception that no appendices are included in the pocket version.

Elias' large-size, general-purpose Arabic-English MODERN DICTIONARY (first published in 1922, Cairo, Egypt) was reported by only 18.5% (N=75 out of 405) of the sample, with nearly half of them (48%, N=36 out of 75) saying they were using it with a 'low' frequency. The entries in this dictionary are arranged according to the Arabic alphabet. The Arabic entries receive pronunciation treatment while their English counterparts do not. No examples are given to clarify meanings, and no grammatical information is provided. The first edition of this dictionary was published in 1922 covering 45,000 words; the number of entries had increased to 63,000 in the fifth edition, 1950.

The main reason why this dictionary was produced was the compiler's discontent with the then available Arabic-English dictionaries which, according to Elias, were designed for orientalists interested in Classical Arabic rather than Arabic-speaking students of English. The preface to this dictionary is in both Arabic and English in which it is mentioned that the dictionary is meant for both Arabic-speaking learners of English and English-speaking learners of Arabic. Obsolete words are excluded and emphasis is placed on 'living words'. Classical as well as colloquial, mainly Egyptian, Arabic is used throughout. The dictionary uses drawings and black-and-white pictures and a thematically-arranged supplement of pictures and drawings is entered (similar to the one in the English-Arabic version).

The POCKET (English-Arabic) MEDICAL DICTIONARY by K.A. Shihabi was reported to be used by 16.8% (N=68 out of 405) of the students, but 42.6% (N=29 out of 68) of them said they 'rarely' used it. The first edition of this dictionary was published in 1964 by Librairie Du Liban, Beirut. Shihabi is a member of staff of the Faculty of Dentistry at the University in Damascus, Syria. He states in the preface (in Arabic) that his dictionary is an outcome of an immediate need to fill the big gap in the Arabic library which lacks a 'modern' medical dictionary giving simplified meanings that are understandable and far from being sophisticated. The target is the medical practitioner in the Arab World. This statement by the compiler confirms a realisation that there is a problem of intelligibility as far as medical Arabic is concerned. However, a close examination of Shihabi's dictionary reveals that it is not very different from HITTI'S MEDICAL DICTIONARY reviewed earlier. It is, in fact, a technical glossary where the Arabic equivalents are in most cases neither explained nor defined. No specific mention is made of the sources on which the POCKET MEDICAL DICTIONARY is based but the compiler mentions the use of medical publications, and translations by staff members of medical faculties at Damascus University, Cairo University, and the American University of Beirut. Like HITTI'S, there is a great deal of space wasted in the body of the dictionary in which phonetic, grammatical, collocational and illustrative materials could have been employed. The dictionary lacks such materials and includes no pictures or drawings save an appendix of 16 anatomical black-and-white plates.

The DICTIONARY OF SCIENTIFIC, TECHNICAL AND ENGINEERING TERMS (English-Arabic, large-size by A. Al-Khateeb, first published by Librairie Du Liban in 1971, Beirut) was reported to be used by 18.5% (N=75 out of 405) of the sample, but out of these students only 20% (N=15 out of 75) indicated a 'high' frequency. The dictionary has 60,000 entries covering various branches of science and technology. In addition, there are nearly 1,300 pictures and illustrative drawings together with many tables showing measurements, units and scientific abbreviations. However, much of the appendixed information is irrelevant to the needs of medical/nursing students, e.g., Geological Time Scale.

The dictionary claims to be comprehensive in that it covers most of theoretical and applied science subjects. The Arabic terminology is based on glossaries published by Arabic Academies in Cairo, Damascus, Baghdad and Rabat. It also refers to general and specialised 'foreign' dictionaries, particularly French and English ones, and there is a bibliography at the end of the dictionary showing the sources. If there is no existing Arabic equivalent, the compiler takes the initiative and creates his own terms taking into account the rules of arabicisation.

The compiler does not specify the usership of this dictionary, but there is a general reference to the needs of the general reader as well as the specialist practitioner. The subject field of each term is indicated through abbreviations and in some cases the part of speech is shown. The entries are arranged strictly alphabetically whether the entry is composed of one word or more, and the dictionary shows both American and British spelling of terms such as gage/gauge.

7.5 Students' Use of Monolingual English Dictionaries

It may be more appropriate to speak of the students' non-use of monolingual English dictionaries as the results indicated that monolingual English dictionaries listed in the questionnaire were reported to be used by fewer students and less frequently in comparison to the reported use of bilingual dictionaries. As Table 12 shows, among the six dictionaries listed, only one pocket technical dictionary can be considered as relatively popular among the sample. This is BAILLIERE'S NURSES' DICTIONARY (BND), by Kasner and Tindall first published in 1912 by Bailière Tindall. More than half (N=90 out of 163) the 40.2% (N=163 out of 405) of the students who reported using it said they were using it with a 'high' frequency. The percentage of users is much higher than 40.2% if we exclude the

Table 12 : Students' use of monolingual English dictionaries

Name of Dictionary	Description		Used by	Frequency low	Frequency average	Frequency high
BAILIERE'S NURSES' DICTIONARY	pocket, technical British	N=	163	27	46	90
		%=	40.2	16.5	28.2	55.2
OXFORD ADVANCED LEARNER'S DICTIONARY OF CURRENT ENGLISH	medium, general purpose British	N=	94	35	33	26
		%=	23.2	37.2	35.1	27.6
LONGMAN DICTIONARY OF CONTEMPORARY ENGLISH	medium, general purpose British	N=	48	25	16	7
		%=	11.8	52.0	33.3	14.5
NEW WEBSTER'S DICTIONARY OF THE ENGLISH LANGUAGE	large, general purpose American	N=	29	19	7	3
		%=	7.1	65.5	24.1	10.3
WEBSTER'S NEW COLLEGIATE DICTIONARY	larger, general purpose American	N=	27	16	5	6
		%=	6.6	59.2	18.5	22.2
COLLINS ENGLISH DICTIONARY	large, general purpose British	N=	22	14	6	2
		%=	5.4	63.6	27.2	9.1

first-year students of whom only 11.8% said they were using it. Both the percentage of users and the percentage of 'high' frequency users seems to increase as the students move up the academic ladder.

During the data-collection period, I observed that some students were carrying this dictionary and when I enquired about it I was told, by a group of fourth-year students, that BND was specifically recommended by the nursing clinical staff. Those students who attend hospital-based training rounds are usually asked to carry a copy of BND during such rounds where almost all communication takes place in English. BND is thought to be helpful for students who may require quick and convenient access to the meaning or pronunciation of a particular medical/nursing term. First-year students are not required to attend such rounds and that may explain why the vast majority of them reported not using it.

BND, according to the preface of its twentieth edition (1984), is intended for "practising nurses and for those employed in related disciplines". However, although it is meant for nursing and medical users, there is no evidence that BND was compiled with the needs of overseas nursing and paramedical students, whose mother tongue is not English, in mind. In fact, BND seems to have been specifically designed for British users; it is introduced by general pages of nursing-related advertisements aimed at British nurses (e.g., nursing careers in the RAF). This should not imply that BND is no good for the ESP nursing learner in general; but the point is rather that despite all its merits, this dictionary remains a reference book that does not seem to take into account the needs of its potential overseas users.

BND was first published in 1912, but it is claimed that it is regularly updated. Pronunciation is indicated, but not by use of the IPA phonetic symbols; it also divides syllables by hyphens and indicates the stressed syllable, which makes pronunciation easy to grasp. The introductory matter includes a pronunciation guide and examples are provided on the pronunciation of consonants and vowels. In addition, irregular plural forms of nouns are given (e.g., acinus, acini). Abbreviations are treated as separate entries and arranged alphabetically. Other features of BND include illustrations distributed throughout the dictionary. The dictionary also relates terms to each other through cross-referencing. As for coverage, there is no indication of how many entries the dictionary includes, but the main body of the dictionary is of 423 (14x 8.5cm) pages. Appended to the dictionary is a 'reference section' of 93 pages; the content of these 21 appendices ranges from prefixes and suffixes to abbreviations showing academic degrees and professional organisations. Although much of the information in the 'reference section' is potentially useful for Arab nursing students (assuming they are capable of understanding it), some of it seems to be applicable to British nurses only (e.g. useful addresses, the Pharmacy and Poisons Act).

As it seems to be aimed primarily at a British audience, one would not expect ESP nursing learners overseas to be able to use it without difficulties in understanding its metalanguage. In fact, inability to comprehend dictionary definitions is often a main reason why non-native learners refrain from using monolingual English dictionaries.

It must be of interest to ESP practitioners as well as pedagogical lexicographers, and dictionary publishers to know that the overwhelming majority of the students in this study did not consider themselves to be users of monolingual English dictionaries including reputable ones for learners such as the OXFORD ADVANCED LEARNER'S DICTIONARY OF CURRENT ENGLISH (OALD) or the LONGMAN DICTIONARY OF CONTEMPORARY ENGLISH (LDOCE); only 23.2% (N=94 out of 405) reported using OALD of whom 27.6% (N=26 out of 94) said they used it with a 'high' frequency, while only 11.8% (N=48 out of 405) reported using LDOCE, more than half (N=25 out of 48) of them said they 'rarely' used it. In her study, El-Badry (in progress) similarly discovered that monolingual English dictionaries (OALD, LDOCE, and WEBSTER'S NEW COLLEGIATE) were said to be used by a very small minority in her sample (under 20%).

Only 4.7% (N=19 out of 405) of the students reported using 'other dictionaries'. Nine of them mentioned they were using a 'medical dictionary' without any specification, while three others said they were using a 'nursing dictionary' but did not specify which. The POCKET MEDICAL DICTIONARY by Nancy Roper, first published in 1933 by Churchill Livingstone, Longman, and the ENCYCLOPAEDIA AMERICANA were reported by the other seven students.

From an ESP perspective, it is somehow discouraging to find out that learners' dictionaries of top quality, such as OALD and LDOCE, are hardly referred to by a group of ESP learners. Publishers, lexicographers and language teaching specialists spend considerable effort, time and money, in providing well-designed dictionaries on the assumption that they will be used by EFL learners somewhere. The findings that the majority of students in this survey reported not using monolingual English dictionaries can be explained with reference to a complexity of factors. These I discuss below.

In Jordan at least, until very recently, monolingual English dictionaries were not recognised as an integral part of the school curriculum and therefore their use was never taught systematically. Pupils naturally resort to their pocket bilingual dictionaries and this develops into a habit which is difficult to give up at later stages. At the same time, little is done by teachers of English to wean these pupils off this habit. In addition, awareness of the role of learners' monolingual English dictionaries in learning English is rather poor among students as well as teachers at the university level. Many students and teachers simply do not know of the existence of monolingual English dictionaries specifically prepared for non-native learners of English as a foreign language.

Moreover, no formal instruction in dictionary reference skills is given at the University of Jordan, and it is left to the individual learner to discover for him/herself reference materials and how to use them. Furthermore, English language teacher-training programmes in Jordan usually lack reference skill components and lexicographical input, and the role of dictionaries is rarely integrated in the curricula.

Other factors pertain to dictionaries rather than users. In some cases, learners' monolingual English dictionaries may not be easily available, and if so they tend to be rather more expensive than their bilingual counterparts. In addition, bilingual dictionaries have the advantage of providing mother-tongue equivalents which are, in the case of general-purpose dictionaries, easier to understand than metalanguage written in English. Although, learners' monolingual English dictionaries do try to simplify their metalanguage, and to use controlled defining vocabulary, as does LDOCE for example, for many students the ability to comprehend readily dictionary definitions is doubtful, due possibly to lack of clarity of dictionary definitions and to the relatively poor general linguistic proficiency on the part of the student.

Finally, we can add that the currently available monolingual English dictionaries tend to be rather sophisticated with so many symbols and abbreviations that, despite helpful introductory guides, they put the learner off. Also, for an ESP student, e.g., a nursing student, much of the content of monolingual English general-purpose dictionaries, at the macro-structure and the micro-structure levels, is irrelevant and redundant. For example, while the punctuation appendix in OALD is useful, the appendices on geographical names and plays of Shakespeare are redundant.

CHAPTER EIGHT
CONCLUSIONS AND RECOMMENDATIONS

8.1 Introduction

This empirical study into aspects of dictionary use among a group of ESP nursing students at the University of Jordan has been carried out from an ESP learning/teaching perspective. It has set out as its major goal to examine the relationship between ESP learners and dictionaries, a theme that has previously received too little attention from researchers in pedagogical lexicography and in ESP teaching. The approach adopted throughout is learner/user-oriented rather than dictionary-oriented. Hence, the bulk of the survey is focussed on the students and their learning environment.

The students provided most of the data through more than one data-gathering instrument, i.e., questionnaires, follow-up interviews and dictionary using records. In addition, the views of the nursing and the ESP teachers were considered via staff questionnaires and interviews. The dictionaries found to be in use were also inspected and critically assessed. A number of theoretical and practical issues have been addressed throughout this work. In this chapter, the overall conclusions that can be drawn are summarised and discussed along with their implications for pedagogical lexicography and ESP teaching. The recommendations for short-term and long-term implementation and suggestions for follow-up research are also presented in this chapter.

8.2 Towards ESP Lexicography

In my view, only an ESP-oriented tool can ideally serve the needs of an ESP customer. Any lexicographical endeavour that does not take into account the underlying principles of ESP cannot, therefore, seriously claim to be appropriate for the interests and the needs of ESP users. In Chapter One, I discussed the notion of ESP and its basic components: the E, the S, and the P. These notions have important implications for pedagogical lexicography. One such implication is that a dictionary that has been compiled before or without the ESP perspective cannot claim professional service to ESP learners.

ESP, as I conceive of it, is essentially pragmatic and therefore lexicography for ESP learners must be pragmatic also. I mean by this, that dictionary design for ESP learners should be bound by the practical needs of the target group of users. It is only when 'purpose specificity', rather than 'language specificity' alone, is employed in lexicography that we can look forward to meeting the needs of the ESP learner as a distinct variety of dictionary user.

The E in ESP represents the English language used in a particular subject context, academic or professional. This E can be spoken as well as written discourse. The implication of this is that a dictionary that is not based on authentic language data of both modes cannot be said to be complete. Many dictionaries tend to exclude spoken English from their sources. In addition, the E of nursing may not be necessarily the same as the E of medicine, though they are obviously inter-related. Therefore, a dictionary that claims to be useful for both subjects is to be treated with caution; (I have, incidently, come across two dictionaries which are almost exactly the same in content, but variously entitled the POCKET MEDICAL DICTIONARY, 13th edition, and the NURSES DICTIONARY, 15th edition, both by the same editor and publisher). Furthermore, the English of nursing should be seen in relation to the communicative contexts in which it is used in a particular academic and/or professional environment. Hence, an ESP dictionary must be based on real linguistic data, spoken and written, that is sampled from the various contexts of language use, e.g., technical reports, textbooks, seminar presentations.

The S of ESP has also important implications for lexicographers. It is no longer a reference to a 'special register' but rather to purposes of learning that have been specified. There are at least two types of purposes in ESP: academic and professional (as far as the learner is concerned), and two types of purposes (as far as the language is concerned): receptive (reading and listening) and productive (writing and speaking). If it is accepted that such purpose types are distinct and involve different language operations, it must follow that a dictionary which claims to fulfil a multiplicity of purposes cannot be properly regarded as an ESP dictionary. I may also add another distinction that can be made between an ESP learner and a translator/interpreter. The implication of this distinction is that a dictionary designed to help professional translators/interpreters may not properly claim efficiency in meeting the particular needs of ESP students. The roles of the two categories of users in using the L are not necessarily similar.

A major principle in ESP is that it is needs-based. Few dictionaries now on the market can claim to have incorporated this principle. Therefore, an ESP dictionary should accommodate insights drawn from needs analyses. Pedagogical lexicography can also learn from ESP the principle of team-teaching/collaboration (the team approach is also a well-established concept in the nursing profession). If dictionaries continue to be compiled by single subject specialists only, I can see little chance that the needs and expectations of ESP learners will be appropriately met. An ESP dictionary should be a well-planned, properly financed project rather than an individual exercise. In such a project, ESP teachers/practitioners, subject teachers, linguistically and pedagogically-oriented lexicographers, publishers, computer personnel, and most importantly, sponsors, should be willing to work in interdisciplinary teams.

Most currently available dictionaries, in my opinion, cannot be taken as reliable ESP dictionaries, despite all their merits. Few of the dictionaries I examined in Chapter Seven can claim to be sufficiently appropriate for the needs of ESP learners. It is true that general learners' monolingual English dictionaries, such as OALD and LDOCE, are to be credited because they aim at satisfying the foreign learner of English; however, they lack specificity as to purposes, particular types of users, and tend to include a lot of materials that are irrelevant to the needs of ESP users as they are meant for every possible learner of English in general and of every cultural background, and for both receptive and productive uses of language. Moreover, because they are general-purpose they cannot cover particular subject terminologies comprehensively, and even if they do, they intend such terminology for the learner as a layman rather than as a student of ESP. Also, such types of dictionaries are becoming increasingly sophisticated and intermediate learners tend to complain about their metalanguage. The ESP learner is to be recognised as a type of user different from the general EFL learner, and therefore needs his/her own type of dictionary.

General-purpose bilingual English-Arabic dictionaries such as AL-MAWRID and Elias' MODERN DICTIONARY are not ESP dictionaries for a variety of reasons. They lack 'purpose-specificity' and fail because they try to meet the needs of diverse categories of users. If ESP students do tend to use such dictionaries, they do so not necessarily because such dictionaries are the best for them but rather because no better alternatives are available. Indeed, despite the fact that the majority of students in this study said they were using AL-MAWRID and that they thought of it positively in comparison to other English-Arabic dictionaries, a majority of 51% against 35% of the sample (Table 87) agreed with the statement that English-English dictionaries were more useful than English-Arabic ones.

Though some of the compilers of the English-Arabic dictionaries discussed in Chapter Seven claim that they designed them with the users' needs in mind, actually none of them refer to the results of any research on language learning, contrastive linguistics, needs analysis or pedagogical lexicography. In addition, while generous in providing redundant information, their treatment of the specialist terminology leaves much to be desired, and the meanings of sub-technical vocabulary are embedded within entries that involve irrelevant senses from the ESP learner's point of view. The medical/nursing terminology entered in popular English-Arabic general-purpose dictionaries is based on ageing English-Arabic medical dictionaries such as HITTI'S MEDICAL DICTIONARY, 1964 and SHARAF MEDICAL DICTIONARY, 1926; on glossaries published by arabicising institutions, and on the compilers' own experiences as translators or subject specialists. As the battle goes on between the 'purists' and the 'innovators' in the Arab World, scientific and technical terminology in bilingual lexicography will continue to suffer from a shortage of terms, a lack of acceptability of new terms, and a dearth of standardisation. Al-Kasimi (1979:112) uses the words 'chaos' and 'confusion' to describe the current situation of terminology in English-Arabic lexicography. Such bilingual technical dictionaries and glossaries, usually aimed at arabicising rather than teaching, cannot be accepted as authoritative ESP dictionaries.

I have found (Chapters Five and Six) that the nursing students needed much help in both decoding and encoding tasks in a variety of communicative contexts. Few bilingual dictionaries, general and technical, can sufficiently help such students in their academic work. Let us take a few examples. AL-MAWRID adopts a historical approach in arranging meanings in entries in order to help the user gain an idea on how human civilisation was developing through history. It is generally observed that learners consider the first sense entered when looking up. Remembering that students do not usually read dictionary front matter, there is a risk of an ESP student taking the oldest meaning of a word as its common meaning. ESP learners are better helped by listing the most common meaning first in an entry, as few of them are likely to have historical interests.

The treatment of meaning of specialist terminology in English-Arabic, general and technical, dictionaries is far from being satisfactory, in my view. Both AL-MAWRID and HITTI'S MEDICAL DICTIONARY provide little help as far as meanings of specialist terminology is concerned. Remembering that medical terms derived from Classical Arabic are not in current use, these dictionaries do little to define or explain specialist terms either in Arabic or in English; they do not provide illustrative examples or pictorial material and they waste much space, by leaving it blank in the case of HITTI'S and by inserting redundant materials in the case of AL-MAWRID.

My needs-analysis has shown that most of the sample students need help in language encoding tasks in a number of academic contexts. Existing English-Arabic technical dictionaries, including HITTI'S, completely ignore the inclusion of phonetic information on English terms (while some provide help in pronunciation of Arabic equivalents). They also lack grammatical and collocational information, and therefore they cannot be assumed to be helpful as far as the skills of speaking and writing are concerned. I can fairly conclude from this that just because a dictionary is about medicine does not necessarily imply that it is an ESP dictionary.

To the best of my knowledge, there are no learners' medical or nursing dictionaries specifically designed for Arab nursing students who study their subjects through English. In the light of my finding that the sample students are currently using dictionaries which are inadequate and incapable of meeting their communicative and lexical needs, and of my argument that an appropriate ESP dictionary does not yet exist, I should like to propose that a genre of ESP dictionaries should be designed to meet the needs of particular groups of ESP users. In particular, I should like to suggest that the University of Jordan take the initiative and consider the feasibility of sponsoring and housing a lexicographical project for ESP purposes targeted at serving ESP users in the Arab World. This is best done in cooperation with universities and Arabic Academies and publishers in the Arab World. Technical expertise may also be sought from the British Council and the American Centre and from Anglo-American universities and publishing houses. Indeed, talks have already taken place concerning a link of linguistic cooperation between the University of Jordan and the University of Birmingham and the possibility of designing an Arabic version of the recently published pioneering COBUILD DICTIONARY has been raised. Practical steps should be taken to maintain this link and the idea of ESP dictionaries for the Arab World should be considered.

As a first step towards introducing ESP dictionaries for Arab users, the findings of this study may be translated into a project of an ESP nursing student dictionary for the Arab World, taking Jordan as a case study. The following guidelines may provide a provisional framework for the dicticnary envisaged:

1. The dictionary should be set up as a project to be carried out by a team of specialists representing all concerned with ESP, i.e., ESP teachers, nursing teachers, lexicographers, publishers, computer specialists, Arabic specialists and sponsors.
2. The target usership should be defined as Arab student nurses who are studying through the medium of English at the tertiary level in an Arabic-speaking learning environment and who are assumed to have studied English as a school subject for 6-8 years.
3. It should be assumed that the target users are not already advanced or proficient in English, in medical Arabic, and in nursing subjects of technical nature. Therefore, the dictionary will not assume that its user needs only labels for known concepts, but explanations and clarifying information that helps him/her understand and use such labels in a learning environment.
4. A computerised linguistic data-base should be set up by sampling authentic spoken and written texts of English used in nursing and nursing-related academic communicative contexts. This will entail access to relatively recent passages of used, or likely to be used, textbooks, handouts, manuals, specialist journals, examination papers, written technical reports and care-plans, written essays and transcripts of recorded lectures, presentations, seminars, oral examinations, and clinical rounds. Such a data-base could also be used for research purposes and the production of classroom materials.

5. The dictionary design should adopt a <u>pragmatic</u> approach rather than a puristic one. Its purpose should be seen as a contribution to the learning/teaching of ESP rather than as a contribution to translation and arabicisation. The approach should also be objective rather than personal through reference to authentic data and expert advice.
6. An ESP dictionary is essentially <u>pedagogical</u> and therefore the proposed dictionary will be assumed to have a teaching role. It should explain and tell the learner how to use English in academic contexts.
7. Ideally, the proposed dictionary should be produced in two versions according to purpose, i.e., decoding and encoding, and in two versions according to size and context of use, i.e., large for use at home or library and pocket size for quick reference use during hospital training and lectures, seminars, and examinations.
8. The decoding version should incorporate more vocabulary items than the encoding one on the basis of the assumption that receptive passive vocabulary is larger than productive active vocabulary. The large-size version should also contain more vocabulary items than the pocket version and include longer entries with more information.
9. The computer should be used to provide concordances showing key words and their textual environments. These can be used as a basis of choosing illustrative examples and determining the context in which a particular entry can be used together with information on its collocations. The computer may also be used to produce statistical data on the frequency of occurrence of particular vocabulary items, a criterion that indicates its currency.
10. The selection of entries should be determined according to the following criteria:
    c. frequency of occurrence

   b. range of occurrence, i.e., texts and contexts in which an item occurs.
   c. usefulness judged from a nursing and ESP point of view.

11. Coverage should include nursing specialist vocabulary (including medical terminology that is judged to be of particular relevance to nursing students), sub-technical vocabulary as understood in ESP, and general but only relevant and necessary vocabulary deemed nursing lexis.
12. The length, size, type of information entered in, and the style/level of explaining the meaning of an entry may not need to be strictly consistent. This should depend on the size of the dictionary, the purpose (encoding or decoding), type of vocabulary item, i.e., an abstract concept may need more careful and longer explanation, with examples, than a term designating a concrete object which can be shown by a picture or an illustration (e.g., an instrument) and a short definition. For the sub-technical and the general but relevant vocabulary items a one-word Arabic equivalent may suffice provided that such equivalent is in common use.
13. The dictionary can be mono/bilingual at the same time. For decoding purposes, it may take the form of three columns in a page starting with the English word on the left, the Arabic explanation or definition in the middle, and the English explanation and other supporting information on the right. For encoding purposes, the page layout may also take three columns: Arabic word (only if likely to be intelligible) on the right, English equivalent in the middle, and information on its meaning, morphology, pronunciation, collocations, usage and how to use it by indicating relevant grammatical information (e.g., parts of speech, irregular plural forms, passivisation, prepositions) and by providing examples and contextual clues.

14. Definitions or explanations should be simplified and the possibility of using the LDOCE's technique of controlled vocabulary should not be ruled out; illustrative examples, preferably authentic ones, and drawings/pictures should be exploited, preferably using colours.
15. In indicating the meanings of sub-technical vocabulary and general relevant nursing vocabulary, only the senses pertaining to the nursing profession should be listed.
16. Guide words should be placed at the top of every page and it is preferred if the key to pronunciation is reproduced at the bottom of every page as in AL-MAWRID.
17. The dictionary should be thumb-indexed to make it easy for the user to locate quickly the needed entry (I saw many students devising their own ways of thumb-indexing their dictionaries; the lexicographer/publisher may provide this feature professionally).
18. The headwords should be bold-face, with dots separating their component syllables. Examples can be italicised. Each different type of information introduced in the entry should be signalled by a symbol or a code. Shading particular information should be considered.
19. Affixes, abbreviations, compounds, and derivatives should be entered as independent entries and listed strictly alphabetically.
20. British as well as American spellings and pronunciations should be indicated and the pronunciation should be in the form of IPA symbols for RP.
21. Lengthy encyclopedic definitions should be generally avoided.
22. Inter-related entries should be cross-referenced throughout.
23. The dictionary may include appendices, but these should have specific pedagogical ESP or nursing functions. In the encoding version, information on irregular verbs, rules of spelling, generally confused lexis etc. may be considered.

24. The introductory matter to the dictionary should be in Arabic and should explain clearly the purposes of the dictionary and how to make best use of it, through explanatory charts and example entries.
25. A booklet, in Arabic, showing, through exercises, how to use the dictionary should accompany the dictionary.

As I stated at the outset of this book, I am an ESP teacher and not too much of a professional lexicographer, and therefore the guidelines proposed above are given by rule of thumb and are open to discussion by the members of the team envisaged, taking into account the practical constraints likely to arise. A more flexible and exhaustive editorial policy would need to be worked out once the idea of the proposed dictionary materialised.

8.3 Pre-University EFL Teaching

In Chapter Four I concerned myself with the situation of dictionaries in the teaching of English at the school level in Jordan. I concluded that the students surveyed in this study could not be assumed to possess any reliable reference skills despite the fact of having learned English, as a school subject, for eight years. The new syllabuses and textbooks, recently introduced at the Compulsory Stage, are certainly a welcome step forward; however, they should be followed up by a move to change teaching patterns at the Secondary Stage so that dictionary awareness could be strengthened and integrated throughout <u>all</u> stages of school learning. By the same token, teacher-training programmes for school teachers of English should ensure that dictionary reference skills are included.

Language teaching planners ought to provide access to class sets of learners' dictionaries. In addition, as it is well-known that learners, particularly at school level, place much value on examinations, and since the current examinations do not include any tasks on dictionary use, examination writers should ensure that examination papers cover this aspect. Although it is generally accepted that bilingual dictionaries are not necessarily harmful, and that it is not always easy to wean students off them, I believe that learners' monolingual English dictionaries can be more useful provided that they are made available and that both learners and teachers are made to realise their value in the learning process.

In this respect, I should like to propose that at the earlier stages of learning English, i.e., classes 5 and 6, the teaching of dictionary use should be based mainly on pocket bilingual dictionaries. At the Preparatory Stage, i.e., classes 7, 8, and 9, bi-/monolingual learners' dictionaries should be introduced, i.e., dictionaries such as AL-MISBAH STUDENTS' DICTIONARY (English-English-Arabic, large-size, by N. Kharma and A. Eyre, first published in 1986 by Librairie du Liban and Longman); while having access to Arabic equivalents, learners may be encouraged to explore the English version in the dictionary. At the Secondary Stage, i.e., classes 10, 11, and 12, full-fledged learners' monolingual English dictionaries, such as the recently launched COBUILD, and the most recent editions of OALD and LDOCE, can be introduced and students should be trained in using them. No attempts should be made to ban bilingual dictionaries in the classroom or in examinations, but students should be shown that learners' monolingual English dictionaries can offer more help. The general principle should be the gradual movement towards learners' monolingual English dictionaries away from bilingual ones.

Most importantly, I should like to recommend that the Ministry of Education consider the feasibility of designing a graded learners' dictionary for Jordanian schools, to accompany and parallel the syllabuses at various stages of learning. The dictionary should be produced in at least three versions: (1) pocket learners' bilingual, (2) medium-size learners' bi-/monolingual, and (3) large-size, intermediate rather than advanced, learners' monolingual English. The use of colour and pictorial materials together with simplified metalanguage, and clear and interesting dictionary exercises should also be envisaged. The first step towards such a move should be a nationwide and comprehensive research project into the current position of dictionaries and the teaching of reference skills at the school level. Furthermore, I should like to suggest that a series of workshops and seminars on the role of dictionary use and the teaching of study/ reference skills should be organised in cooperation with the universities with the aim of increasing the awareness of reference tools among in-service teachers of English in the country. Such measures would help to produce pupils who can join tertiary and further education with confidence and a sense of independence in acquiring knowledge and skill in their chosen subjects of study.

8.4 University EFL/ESP Teaching and Teacher-Training

In Chapter Five, I found that none of the compulsory University courses, i.e., 'English 99', and 'English 100' had any lexicographical input. Therefore, I propose that such courses should include a specific component on dictionary reference skills. The list of objectives in the design document for each course should state that students by the end of the course should be able to use a learners' monolingual English dictionary proficiently. The design document should also include a syllabus of the proposed reference skills component, and a set of master copies of materials to which the teachers can refer for classroom use.

A committee should be set up within the English Section of the Language Centre to monitor the writing of the proposed syllabus and the accompanying materials. In addition, I recommend that class sets of recent editions of learners' monolingual English dictionaries are made available for classroom use. Even if it is agreed that the students taking 'English 99' and 'English 100' receive instruction in dictionary use, I believe that this will not be taken seriously as long as the University Proficiency Examination (UPE) does not include questions on this aspect. Recently, the UPE has been modified to be administered weekly and for small groups of students. This should make it possible, in principle, for the examiners to watch students performing dictionary-using tasks, e.g., locating an entry quickly; written questions could also be included in the UPE. I should also like to suggest that any teachers responsible for the teaching and testing of the dictionary input in the compulsory courses and the UPE should attend a minimum number of special workshops/ seminars in order to familiarise themselves with the value of the new component in their teaching.

Of particular relevance to this study are the courses 'Advanced English Vocabulary' and 'Study Skills in English' in the University Elective Programme. Both courses mention the use of dictionary and reference materials in their course descriptions. As the whole programme is new, and optional, students and staff members in all departments should be made aware of the existence and value of such courses. Pamphlets introducing these courses should be designed and circulated and the Language Centre staff should start preparing the course design documents with specifications on syllabuses, materials, methods, and tests. The other four courses in the programme should also integrate dictionary input into all aspects of their design so that dictionary use becomes part and parcel of the learning of the four language skills emphasised in such courses.

As a graduate of the University of Jordan Department of English, I could not fail to notice that there is no course on lexicography as such and students are hardly trained at all in the use of dictionaries and reference materials. I therefore urge the English Department to consider the incorporation of pedagogical lexicography in its study plans as a component of both undergraduate and postgraduate programmes. It is also suggested that teacher-trainees of English in the Faculty of Education receive instruction in how to use dictionaries and how to teach dictionary use to their future students. The 'word-card-dictionary' approach seems interesting and therefore should continue to be part of the 'Vocabulary' course, but I suggest that more than 5% of the final grade should be allocated to this component. In addition, it is not enough to 'encourage' students to buy a 'good' dictionary; I think students should be

<u>asked</u> to purchase a learners' monolingual English dictionary and should be given direct help in making their choices.

The Language Centre is now a major partner in the training of EFL/ESP teachers; however, I found that none of its present teacher-training courses provides any lexicographical input. I also noticed that the awareness of the teacher trainees in the programme left much to be desired. I therefore wish to put forward the following proposals:

1. The courses on 'Language Skills: 1' and 'Language Skills: 2' should include reference skills to increase the trainees' own knowledge and skills in using dictionaries and other reference tools. The design document in each course should mention clearly the ability to use dictionaries and reference works in the list of objectives and the course description. The teaching and testing materials should include exercises and tasks on various aspects of reference skills. The first steps have already been taken in this direction as reflected in an interesting experiment recently carried out by the ex-Teacher-Training Adviser and myself.
2. The two undergraduate courses on 'Methods of Teaching EFL: 1' and 'Methods of Teaching EFL: 2' should be amended to accommodate a component on methodology of teaching reference skills and on the role of dictionaries in the learning of EFL.
3. There is some scope for a lexicographical input in the course on 'Language Variation'. The syllabus may be modified to include the role of the dictionary in recording language variation. The notion of dictionary 'authority' could be discussed in relation to the role of the dictionary in prescribing and/or describing language. The course may also mention the labelling of style and usage in pedagogical lexicography and how students could be trained to make most use of such information in dictionaries.

4. Of special relevance to this research is the course on 'Teaching of ESP'. The design document should be re-written in order to allow for the integration of topics in pedagogical lexicography. Students taking this course should be expected to receive tuition on the role of dictionaries in ESP. The course should aim at developing the trainees' abilities to analyse and evaluate dictionaries as part of their teaching.
5. Part of the course on 'Language Description' should look at the implications of describing language in dictionary-making. The syllabus could incorporate an element in the relationship between linguistics and pedagogical lexicography and the problems involved.
6. The possibility of introducing an independent course on 'Dictionaries and Vocabulary in EFL' should be considered, taking into account the pressure on the study plan and the time-table.
7. The back-up materials in the Resource Collection were found to include few lexicographical materials. I suggest that sets of dictionaries of various types, as well as books on pedagogical lexicography should be selected and purchased; and that the University Library be asked to stock books on lexicography and subscribe to journals on lexicography such as the recently launched International Journal of Lexicography, published by Oxford University Press.

As far as the ESP nursing programme is concerned, I suggest that a number of seminars and special workshops on dictionary evaluation for ESP purposes should be held as part of the Language Centre's weekly academic meetings. The role of dictionaries in ESP learning should be discussed and ways of enhancing dictionary skills among the ESP students should be considered. The ESP teachers should also be helped in gaining access to dictionaries and guidance to the literature on lexicography. I say this because I believe there is little hope in helping the ESP students if we do not first develop the ESP teachers' own lexicographical awareness. My interviews with the six ESP teachers have shown that only two of them said they had received any lexicographical instruction during their undergraduate and postgraduate studies and that such input had been inadequate. The ESP teachers also said they rarely carried dictionaries to their classes; and while half of them said they had no idea which dictionaries their students were using the other half could only guess. Although five of the teachers said they usually encouraged their students to use dictionaries, they reported giving little help in dictionary use as part of their teaching, and moreover, they found it hard to recommend dictionaries for their students because they themselves admitted a lack of dictionary awareness.

As for the nursing students themselves, there is evidence, as shown in Chapter Four, that they join the University without having been exposed to sufficient training in reference skills in general, and in dictionary use in particular. Bearing this in mind, I should like to recommend that the Language Centre set up a committee with the brief of redesigning the ESP nursing programme in the light of the findings I have reported. This committee may wish to consider the following guidelines, aiming to integrate reference skills into the ESP course design:

1. The current placement testing system should be expanded to allow for some questions on proficiency in dictionary skills.
2. A syllabus of dictionary skills should be written first and then distributed over the four courses on offer, taking into account that each course has a focus on a particular language skill, e.g., report writing.
3. Before a detailed syllabus on dictionary skills is compiled, the list of course objectives and the course description in each course should be altered to include a clearly-stated item mentioning what students should be expected to know/do concerning dictionaries by the end of each course.
4. The design of the proposed syllabus for dictionary skills should preferably focus on three dimensions: knowledge, skills, and attitudes; the syllabus should

aim at orientating students towards a more positive view of the dictionary role in their learning.

5. The proposed syllabus should mirror dictionary contents and formats, macro- and micro-structures; it should also cover various types of dictionaries that students may need to consult. Provisionally, I should like to propose the following list, which is by no means exhaustive, to be considered for inclusion in the suggested syllabus:
   a. What is a dictionary?
   b. Types of dictionaries.
   c. Buying/choosing a dictionary.
   d. The difference between dictionaries and other reference tools.
   e. When to use and when not to use a dictionary.
   f. Information a dictionary can provide and how to access it: semantic,phonetic, morphological, orthographic, grammatical, stylistic, etymological, cultural, encyclopedic.
   g. Dictionary use for receptive purposes: reading and listening.
   h. Dictionary use for productive purposes: speaking and writing.
   i. Dictionary introductions/guides: task-based discussions of contents and formats of dictionaries in current use.
   j. Metalanguage in dictionaries, i.e., various terms usually employed in talking about dictionaries students may not already know such as: cross-referencing, definition/ sense, usage label, stylistic information, collocation, inflections, entry.
   k. Locating an entry.
   l. Speed in the looking-up process.
   m. Skimming and scanning an entry.
   n. Intelligent guessing and the use of contextual clues and the role of the dictionary in verifying guessing.

6. The faculty of Nursing should consider the possibility of recognising the ESP courses as fully-credited. The proposed re-design is not likely to be effective as long as the present pass/fail status continues.
7. It is imperative that the re-design includes an item on book and library awareness with the possibility of arranging guided visits to the Library and showing students how to make maximum use of its facilities, including reference materials.
8. The University of Jordan's Bookshop has to be notified which dictionaries and other reference works are needed and arrangements should be made to provide the Language Centre with class sets and inspection copies of recent editions of dictionaries while students should be made aware of the availability of reference tools in the Bookshop.
9. The nursing students appreciate relevance of materials to their subject-matter. Therefore, it is suggested that the course materials, including the proposed integrated lexicographical component, should be based on texts as relevant to nursing as possible.
10. I suggest that dictionary use should not be banned in examinations and that a dictionary in examinations, both nursing and language, should be made optional. Examination papers should avoid questions to which a dictionary can provide straightforward answers.
11. Every effort should be made to familiarise the students with the fact that dictionaries, particularly learners' monolingual English ones, can and do have a positive role in helping writing and speaking. Students should be shown how they can make use of reference tools in productive language-use contexts.
12. To encourage students to use dictionaries and realise their value, ESP and nursing staff should consider the possibility of assigning tasks that demand the consultation of a dictionary. The teachers should also minimise their own role as providers of lexical in-

formation so that students become more independent in their learning.

13. The design of the lexicographical component should emphasise the need to help students gradually transfer from reliance on unreliable bilingual dictionaries to learners' monolingual English ones. At present, I am aware of no existing learners' monolingual English medical or nursing dictionaries, but students can benefit from training in the use of the currently available learners' monolingual English dictionaries such as OALD, LDOCE and COBUILD, and existing medical/ nursing dictionaries.
14. Teaching materials for the proposed lexicographical input should be compiled, using a variety of sources. At present, exercises/ tasks on dictionary use can be found in dictionary-using guides designed to accompany particular dictionaries, such as 'Use Your Dictionary' by A. Underhill (1980), 'Learning with LDOCE' by J. Whitcut (1979), and 'How to Use Your Dictionary' by R. Lewis and M. Pugmire (1980). In addition, there are exercises on dictionary use sometimes inserted in vocabulary sections in EFL textbooks. The ESP staff may also find it necessary to design their own materials depending on the particular objectives they want to achieve.
15. Teaching the lexicographical component in ESP courses should make use of the available audio-visual aids, e.g., tape recorders, overhead projectors and language laboratories. Dictionary use should be taught as tasks to be carried out, in class and as homework, through pair and group-work where students can do things rather than just listen passively to lectures on dictionaries. To make teaching/ learning dictionary skills interesting and effective, one can think of ways of exploring the use of techniques such as the slide-tape programme suggested by Clarke and Upton (1982).

8.5 The Need for Further Research

The relationship between ESP and lexicography is a wide-ranging topic involving many issues, and therefore a single study like this can only claim to be a modest start. ESP is still relatively a new trend in language teaching and research, and particularly with regard to vocabulary and reference skills, it remains seriously insufficient (cf. Swales 1983, 1985b). It is only recently that pedagogical lexicography and the user-perspective have gained momentum, but from my literature search I came to realise that the ESP learner as a dictionary user, in the Arab World in particular, has received only scant attention from researchers.

An important aim of this study has been to raise the issue and invite other researchers to follow up this work with similar studies on their own ESP situations so that a comprehensive profile can be built up. Because of the constraints of time and other practical matters, the scope of this survey had to be confined to cover only one group of ESP learners, i.e., Jordanian student nurses. However, it offers a case for comparison. Similar projects should be initiated to survey the needs of ESP learners in Jordan and in other Arab countries so that lexicographers, teachers, and publishers orientate their input on the basis of objective data rather than intuition.

An area of enquiry may be a comparison between dictionary-using strategies among native English-speaking science and technology students and dictionary-using strategies among non-native foreign learners of ESP. Just how ESP learners actually go about dictionary use is not yet clear as most research is based on indirect data-collection methods, and therefore I suggest that more thought should be given to experimental and 'protocol' methods.

Intelligibility of dictionary entries is a potential area for research. The linguistic difficulties met by ESP learners while interacting with an entry text may be clarified. The typology of words frequently looked up by ESP learners may be investigated and features of such lexis may be used for dictionary and syllabus design. In addition, experimental work may be undertaken to discover improved and even radical ways of presenting lexicographical data in both paper and computerised dictionaries. Furthermore, thought may be given to thematic and pictorial lexicography and its potential in helping ESP learners.

Researchers may also carry out linguistic analyses on nursing, or any other variety of English, with the aim of providing ESP teachers and lexicographers with insights into the nursing lexis. Are there fixed expressions and collocations in súch lexis? Can such lexis be classified according to notions/functions and/ or genres? Studies on error analysis in relation to particular contexts of ESP may also provide ESP lexicographers with useful insights. More research is also needed on attitudes of ESP students, ESP teachers and teacher-trainees towards various aspects of dictionary design and dictionary use. The effectiveness of particular dictionaries in the learning of ESP in a specific situation can be measured experimentally, and through tests and surveys.

Also, it will be interesting to explore the effectiveness of the proposed integration of lexicography in the ESP design through studying teachers' and students' attitudes towards the new component and through testing their performance after tuition. I know of little research work on criteria for dictionary evaluation for pedagogical purposes or on criteria for materials design on dictionary skills. Furthermore, we know little about the methodology of teaching dictionary skills; experimental research can produce useful results in this respect.

Finally, this study could not cover dictionaries and ESP students outside the field of nursing. There is scope for studying technical dictionaries and glossaries, monolingual and bilingual, and their use by students in other subjects such as engineering and law.

BIBLIOGRAPHY

References: Dictionaries

A. Monolingual English Dictionaries:

BAILLIERE'S NURSES' DICTIONARY, (1912). 20th edition (1984) (1986 imp.), Kasner and Tindall, Eastbourne, East Sussex: Bailière Tindall.

CHAMBERS UNIVERSAL LEARNERS' DICTIONARY. (1980), E.M. Kirpatrick, Edinburgh: Chambers.

COLLINS COBUILD ENGLISH LANGUAGE DICTIONARY, (1987), J. Sinclair, London and Glasgow: Collins.

COLLINS ENGLISH DICTIONARY, (1979), P. Hanks et al. London and Glasgow: Collins.

CONCISE MEDICAL DICTIONARY, (1980), 2nd edition (1985), E.A. Martin, Oxford University Press.

A CONCISE PRONUNCIATION DICTIONARY OF BRITISH AND AMERICAN ENGLISH, (1972), J. Windsor Lewis, Oxford University Press.

A DICTIONARY OF ENLGISH SYNONYMS, (1986), (1987 imp.), R. Soule, Ware, Hertforshire: Omega.

DORLAND'S POCKET MEDICAL DICTIONARY, (1983), 23rd edition (1982), K. McCullough, Philadelphia: Saunders.

THE EVERYMAN ROGET'S THESAURUS OF ENGLISH WORDS AND PHRASES, 1986 (1987 imp.), D.C. Browning, London: Chancellor Press.

LEARNER'S DICTIONARY OF SCIENCE AND TECHNOLOGY, (1983), C. Lacey et al., London: Bell and Hyman.

LONGMAN DICTIONARY OF APPLIED LINGUISTICS, (1985), J. Richards, J. Platt and H. Weber, London: Longman.

LONGMAN DICTIONARY OF CONTEMPORARY ENGLISH, (1978) (1984 imp.), P. Proctor et al., Harlow and London: Longman.

LONGMAN DICTIONARY OF THE ENGLISH LANGUAGE, (1984) (1985 imp.), H. Gay et al., Harlow: Longman.

LONGMAN DICTIONARY OF SCIENTIFIC USAGE, (1979) (1981 imp.), A. Godman and E.M.F. Payne, Harlow and London: Longman.

LONGMAN LEXICON OF CONTEMPORARY ENGLISH, (1981) (1985 imp.), T. McArthur, Harlow and London: Longman.

NURSES DICTIONARY, 1932, 15th edition (1978) (1985 imp.), N. Roper, Edinburgh, London and New York: Churchill Livingstone.

OXFORD ADVANCED LEARNER'S DICTIONARY OF CURRENT ENGLISH, (1948), 3rd edition (1974) (1985 imp.), A.S. Hornby et al., Oxford University Press.

THE OXFORD-DUDEN PICTORIAL ENGLISH DICTIONARY: SCIENCE AND MEDICINE, (1986), Oxford University Press.

THE PENGUIN WORDMASTER DICTIONARY, (1987), M.H. Manser, Harmondsworth: Penguin.

POCKET DICTIONARY FOR NURSES, (1984), P. Wainwright et al., Oxford University Press.

POCKET MEDICAL DICTIONARY, (1933), 13th edition (1978), N. Roper, Edinburgh, London and New York: Churchill Livingstone.

THE POCKET OXFORD DICTIONARY OF CURRENT ENGLISH, (1924), 7th edition (1984) (1985 imp.), R.E. Allen, Oxford University Press.

TABER'S CYCLOPEDIC MEDICAL DICTIONARY, (1940), 11th edition (1970), C.W. Taber, Philadelphia: Davis.

WEBSTER'S NEW COLLEGIATE DICTIONARY, (1979) 8th edition, H.B. Woolf et al., Springfield, Massachusetts: Merriam.

B. <u>Bilingual Dictionaries (English-Arabic)</u>

AL-MAWRID: A Modern English-Arabic Dictionary, (1967), 20th edition (1986), M. Ba'albaki, Beirut: Dar El-Ilm Lil-Malayen.

AL-MAWRID AL-QUAREB: Ba'albaki's Pocket Dictionary, (1968) (1980 imp.), M. Ba'albaki, Beirut: Dar El-Ilm Lil-Malayen.

AL-MAWRID AL-WASET: A Concise English-Arabic Dictionary, (1971), 3rd edition (1977), M. Ba'albaki, Beirut: Dar El-Ilm Lil-Malayen.

AL-MISBAH STUDENTS' DICTIONARY: English-English-Arabic, (1986), N. Kharma and A. Eyre, Harlow: Longman and Beirut: Librairie du Liban.

AL-MUFID: A Learner's English-Arabic Dictionary, (1985), R. Nasr and A. Al-Khatib, Beirut: Librairie du Liban.

AL-MUGHNI AL-AKBAR, (1987), H.S. Karmi, Beirut: Librairie du Liban.

DICTIONARY OF SCIENTIC, TECHNICAL AND ENGINEERING TERMS, (1971) 6th edition (1984), A. Al-Khatib, Beirut: Librairie du Liban.

Elias' MODERN DICTIONARY, (1913), 23rd edition (1979) (1986 imp.), E.A. Elias and E.E. Elias, Cairo: Elias' Modern Publishing House.

Elias' POCKET DICTIONARY, (1973) (1986 imp.), E.A. Elias and E.E. Elias, Cairo: Elias' Modern Publishing House.

AN ENGLISH-ARABIC DICTIONARY OF MEDICINE, BIOLOGY, AND ALLIED SCIENCES, (1929), M. Sharaf, Cairo: Government Press.

AN ENGLISH-ARABIC READER'S DICTIONARY, (1974) (1980 imp.), J. John Abcarins, Beirut: Librairie du Liban.

HARRAP'S ENGLISH-ARABIC BASIC LEARNER'S DICTIONARY, (1986), N. Abboud, London: Harrap.

HITTI'S MEDICAL DICTIONARY, (1967), 4th edition (1982) (1984 imp.), Y. Hitti, Beirut: Librairie du Liban.

A LEARNER'S ENGLISH-ARABIC DICTIONARY, (1984 imp.), F. Steingass, Beirut: Librairie du Liban.

MEDICAL TERMS DICTIONARY, (1978), K.E. El-Hennawi, Beirut: Modern Library.

THE OXFORD ENGISH-ARABIC DICTIONARY OF CURRENT USAGE, (1972) (1979 imp.), N. Doniach, Oxford: Clarendon Press.

A POCKET FIRST ENGLISH-ARABIC DICTIONARY, (1985), A. Frisby and A. Al-Khatib, Beirut: Librairie du Liban.

POCKET MEDICAL DICTIONARY, (1964), 2nd edition, (1983), K.A. Shihabi, Beirut: Librairie du Liban.

OTHER REFERENCES

Abd-EL-Jawad, H. (1986) 'A linguistic analysis of spelling errors made by Jordanian university students'. Abhath AL-Yarmouk, Vol. 4, No.1, pp. 19-31, Yarmouk University, Jordan.

Abercrombie, D. (1978) 'The indication of pronunciation in reference works', in P. Strevens (ed.), pp. 119-126.

Adams-Smith, D. (1979) 'The acquisition of technical vocabulary', ESPMENA Bulletin, Vol. 1, No. 10, pp. 18-25.

Adams-Smith, D. (1983) 'Style in medical journals', British Medical Journal, No. 287, pp. 1122-1124.

Al-Besbasi, I. (in progress) Translation Equivalence Strategies, Including the Use of Dictionaries, Ph.D. thesis, University of Exeter, (provisional title).

Al-Chalabi, S. (1984) 'Modern Arabic terminology and bilingual lexicography: activities and problems', in J. Swales and H. Mustafa (eds.), pp. 280-291.

Al-Kasimi, A. (1977) Linguistics and Bilingual Dictionaries, Leiden: Brill (1983 imp.).

Al-Kasimi, A. (1979) 'Problems of technical terminology in Arabic lexicography', in R. Hartmann (ed.) (1979a), pp. 111-118.

Allwright, J. and R. Allwright (1977) 'An approach to the teaching of medical English', in S. Holden (ed.), pp. 58-62.

Anderson, J. (1980) 'The lexical difficulties of English medical discourse for Egyptian students', English for Specific Purposes Newsletter, No. 37, Oregon State University, p. 4.

Andrews, S. (1984) 'The effect of Arabicisation on the role of service English', in J. Swales and H. Mustafa (eds.), pp. 172-183.

Ard, J. (1982) 'The use of bilingual dictionaries by ESL students while writing', ITL: Review of Applied Linguistics No. 58, pp. 1-27.

Atkins, B. (1985) 'Monolingual and bilingual learners' dictionaries: a comparison', in R. Ilson (ed.), pp. 15-24.

Atkins, B., H. Lewis, D. Summers and J. Whitcut (1987), 'A Research Project into the Use of Learners' Dictionaries', in A. Cowie (ed.) (1987b), pp. 29-43.

Austin, J. (1962) How to Do Things with Words, Oxford: Clarendon.

Ayto, J. (1983) 'Semantic analysis and dictionary definitions', in R. Hartmann (ed.) (1983a), pp. 89-98.

Aziz, Y. (1980) 'Some problems of English word-stress for the Iraqi learner', ELT Journal, Vol. 34, No. 2, pp. 104-109.

Baker, M. (1987) A Design for a Bilingual English-Arabic Medical Dictionary, M.A. dissertation, University of Birmingham.

Baker, M. (1988) 'Sub-technical vocabulary and the ESP teacher: an analysis of some rhetorical items in medical journal articles', Reading in a Foreign Language, Vol. 4, No. 2, pp. 91-105.

Balet, S. (1984) 'Getting the students to talk about themselves', Modern English Teacher, Vol. 12, No. 2, pp. 12-13.

Barber, C. (1962) 'Some measurable characteristics of modern scientific prose', in J. Swales (ed.) (1985b), pp. 1-16.

Barnhart, C. (1962) 'Problems in editing commercial monolingual dictionaries', in F. Householder and S. Saporta (eds.), pp. 161-181.

Barnhart, C. (1980) 'What makes a dictionary authoritative', in L. Zgusta (ed.), Theory and Method in Lexicography: Western and Non-Western Perspectives, Columbia, South Carolina: Hornbeam Press, pp. 33-42.

Bauer, L. (1983) English Word-Formation, Cambridge University Press.

Baxter, J. (1980) 'The dictionary and vocabulary behaviour: a single word or a handful?', TESOL Quarterly, Vol. 14, No. 3, pp.325-336.

Beatie, N. (1973) 'Teaching dictionary use', Modern Languages, Vol. 54, No, 4, pp. 161-167.

Béjoint, H. (1981) 'The foreign student's use of monolingual English dictionaries: a study of language needs and reference skills', Applied Linguistics, Vol. 2, No. 3, pp. 207-222.

Béjoint, H. and A. Moulin (1987) 'The place of the dictionary in an EFL programme', in A. Cowie (ed.) (1987b), pp. 97-114.

Bensoussan, M., D. Sim and R. Weiss (1981) 'The effect of dictionary usage on EFL test performance compared with student and teacher attitudes and expectations', in B. Sigurd and J. Svartik (eds.), AILA 81, Proceedings 1, Lund University, pp. 216-217.

Bensoussan, M. (1983) 'Dictionaries and tests of EFL reading comprehension', ELT Journal, Vol. 37, No. 4, pp. 341-345.

Bensoussan, M. and B. Laufer (1984) 'Lexical guessing in context in EFL reading comprehension', Journal of Research in Reading, Vol. 7, No. 1, pp. 15-32.

Bird, N. (1984) 'Etymology and the teaching of vocabulary to medical students', English for Specific Purposes Newsletter, No. 9, Oregon State University, pp. 1-3.

Bonamy, D. and J. Moon (1981) 'Some strategies for training ESP teachers overseas', RELC Journal, Vol. 2, No. 2, pp. 76-96.

Boyle, J. (1980) ' The strategies for improving lexical proficiency in ESP', Indian Journal of Applied Linguistics, Vol. 6, No. 1, pp. 86-93.

Bramki, D. (1983) A Study of Lexical Familiarization in Economic Texts, M.Sc. dissertation, University of Aston in Birmingham.

Bramki, D. and R. Williams (1984) 'Lexical familiarization in economic text, and its pedagogic implications in reading comprehension', Reading in a Foreign Language, Vol. 2, No. 1, pp. 169-181.

Brelsforth, P. (1982) Oral and Non-Oral Patterns of Interaction in the Nursing Process, with Implications for English Course Design for Nurses, M.Sc. dissertation, University of Aston in Birmingham.

Brumfit, C. (1985) 'Preface', in R. Ilson (ed.), p. v.

Bujas, Z. (1975) 'Testing the performance of a bilingual dictionary based on topical current texts', Studia Romanica et Anglica Zagrabiensia, No. 39, pp. 193-204.

Burton, J. (1988) 'I want to learn "real" English: lexical links between theory and practice in teaching ESL', in T. Burton and J. Burton (eds.), pp. 163-175.

Burton, T. and J. Burton (eds.) (1988), Lexicographical and Linguistic Studies: Essays in Honour of G.W. Turner, Cambridge: Brewer.

Butler, C. (1985) Statistics in Linguistics, Oxford: Blackwell.

Bygate, M. (1987) Speaking, Oxford University Press.

Byren, D. (1979) Teaching Writing Skills, London: Longman.

Calzolari, N., E. Picchi and A. Zampolli (1987) 'The use of computer in lexicography and lexicology', in A. Cowie (ed.) (1987b), pp. 55-77.

Carter, R. (1987) Vocabulary: Applied Linguistic Perspectives, London: Allen and Unwin.

Carter, R. and M. McCarthy (eds.) (1988) Vocabulary and Language Teaching, London: Longman.

Chabner, D. (1985) The Language of Medicine, Philadelphia and London: Saunders

Chagunda, M. (1983) Interfaces between Technical Dictionaries and EAP Learners, M.Sc. dissertation, University of Aston in Birmingham.

Chamberlain, R. and M. Flanagan (1978) 'Developing a flexible ESP programme design' English for Specific Purposes, (ELT documents 101), London: The British Council, pp. 36-55.

Chambers, F. (1980) 'A re-evaluation of needs analysis in ESP', ESP Journal, Vol. 1, No. 1, pp. 25-33.

Clarke, D. and I. Nation (1980) 'Guessing the meaning of words from context: strategy and techniques', System, Vol. 8, pp. 211-220.

Clarke, N. and C. Upton (1982) 'A slide-tape programme for teaching dictionary skills', RELC Guidelines No. 4, pp. 82-88.

Coffey, B. (1984) 'ESP - English for Specific Purposes', Language Teaching, Vol. 17, No. 1, pp. 2-16.

Cohen, A., H. Glasman, P. Rosenbaum-Cohen, J. Ferrara, and J. Fine (1988) 'Reading English for specialized purposes: discourse analysis and the use of student informants', in P. Carrell, J. Devine and D. Eskey (eds.) Interactive Approaches to Second Language Reading, Cambridge University Press, pp. 152-167.

Cohen, L. and L. Manion (1985) Research Methods in Education, London: Croom Helm.

Connolly, P. (1973) 'How to teach families of words by comparison', ELT Journal, Vol. 27, No. 2, pp. 171-176.

Cowie, A. (1978a) 'The place of illustrative material and collocations in the design of a learner's dictionary', in P. Strevens (ed.), pp. 127-139.

Cowie, A. (1978b) 'Vocabulary exercises within an individualised study programme', Individualisation in Language Teaching, (ELT documents: 103), London: The British Council, pp. 37-45.

Cowie, A. (1979) 'The treatment of polysemy in the design of a learner's dictionary', in R. Hartmann (ed.) (1979a), pp. 82-88.

Cowie, A. (1981a) 'Lexicography and its pedagogic applications: an introduction', Applied Linguistics Vol. 2, No. 3, pp. 203-206.

Cowie, A. (1981b) 'The treatment of collocations and idioms in learners' dictionaries', Applied Linguistics, Vol. 2, No. 3, pp. 223-235.

Cowie, A. (1983a) 'English dictionaries for the foreign learner', in R. Hartmann (ed.) (1983a), pp. 135-144.

Cowie, A. (1983b) 'On specifying grammatical form and function', in R. Hartmann (ed.) (1983a), pp. 99-107.

Cowie, A. (1984) 'EFL dictionaries: past achievements and present needs', in R. Hartmann (ed.), pp. 155-164.

Cowie, A. (1987a) 'Syntax, the dictionary and the learner's communicative needs', in A. Cowie (ed.) (1987b), pp. 183-192.

Cowie, A. (ed.) (1987b) The Dictionary and the Language Learner, Tübingen: Niemeyer

Cowie, A. (1988) 'Stable and creative aspects of vocabulary use', in R. Carter and M. McCarthy (eds.), pp. 126-139.

Crocker, A. (1981) The Development of a University Service English Programme (English Language Working Paper 1), unpublished report, Amman: The Language Centre, Faculty of Arts, University of Jordan.

Crocker, A. (1982) 'Skill and information in university service English courses: experience at the University of Jordan Language Centre', paper presented to the conference on 'Problems of Teaching English Language and Literature at Arab Universities', The University of Jordan.

Crocker, A. (1983) Language Centre Research Report 1: Student Attitudes towards English Language Learning, unpublished manuscript, The University of Jordan.

Crocker, A. (1984) 'Method as input and product of LSP course design', in J. Swales and H. Mustafa (eds.), pp. 129-150.

Crocker, A. (1985) University of Jordan Language Centre, English Section, Guide to Programmes, unpublished manuscript.

Crystal, D. (1971) Linguistics, Harmondsworth: Penguin.

Crystal, D. (1981) Directions in Applied Linguistics, London: Academic Press.

Cucinella, B. (1984) 'Teaching listening to university students', Modern English Teacher, Vol. 12, No. 2, pp. 32-33.

Cunningsworth, A. (1983) 'Needs analysis - a review of the state of the art', System, Vol. 11, No. 2, pp. 149-154.

Dau'ud, A. (1984) Analysis of the Errors Belonging to the English Prepositional System Made by Jordanian Secondary School Students, M.Ed. dissertation, University of Jordan.

Davies, P. (1985) Medical Terminology: A Guide to Current Usage, London: Heinemann Medical.

Diab, T. (1983) Linguistic Analysis as an Input to the Thinking Behind ESP Materials Selection for Medical and Nursing Students with Particular Reference to the Jordanian Situation, M.Sc. dissertation, University of Aston in Birmingham.

Diab, T. (1984) 'Linguistic analysis as an input to course design with special reference to nursing English', paper presented at the 2nd Linguistics Conference of Yarmouk University.

Diab, T. (1988) 'A purpose-specific approach to pedagogical lexicography', paper presented at BAAL Annual Conference on 'WORDS', University of Exeter.

Dirckx, J. (1987) ' The technical language of medicine', EMP Newsletter, (Kuwait University) Vol. 4, No. 1, pp. 38-42.

Drysdale, P. (1987) 'The role of examples in a learner's dictionary', in A. Cowie (ed.) (1987b), pp. 213-224.

Dubois, J. (1981) 'Models of the dictionary: evolution in dictionary design', Applied Linguistics, Vol. 2, No. 3, pp. 236-249.

Dudley-Evans, T. and J. Swales (1980) 'Study modes and students from the Middle East', Study Modes and Academic Development of Overseas Students: (ELT documents: 109), London: The British Council, pp. 91-101.

El-Badry, N. (1986) 'The development of the bilingual English-Arabic dictionary from the middle of the nineteenth century to the present', in R. Hartmann (ed.) (1986b), pp. 57-64.

El-Badry, N. (in progress) Bilingual Learners' Dictionaries for Arab University Students of English, Ph.D. thesis, University of Exeter (provisional title).

El-Halees. Y. (1986) 'The teaching-learning situation of English pronunciation in Jordan', System, Vol. 14, No. 1, pp. 59-70.

Ellegård, A. (1978) 'On dictionaries for foreign learners', Moderna Språk, Vol. 72, No. 3, pp. 225-242.

Ellis, R. (1986) 'Activities and procedures for teacher training', ELT Journal, Vol. 40, No. 2, pp. 91-99.

EL-Sakran, T. (1985) The Treatment of Phrasal and Prepositional Verbs in General Purpose English-Arabic Dictionaries, M.Sc. dissertation, University of Aston in Birmingham.

Escorcia, B. (1985) 'ESP and beyond: a quest for relevance', in R. Quirk and H. Widdowson (eds.) English in the World: Teaching and Learning the Language and Literatures, Cambridge University Press, pp. 228-244.

Ewer, J. (1983) 'Teacher-training for EST: problems and methods', ESP Journal, Vol. 2, No. 7, pp. 9-31.

Faerch, C., K. Haastrup and R. Phillipson (1984) Learner Language and Language Learning, Clevedon: Multilingual Matters.

Fareh, S. (1984) How Far Does the English Language Teachers' Pronunciation Affect that of their Students in Jordanian Public Secondary Schools?, M.Ed. dissertation, University of Jordan.

Farrel, D. and M. St John (eds.) (1987) The ESP Reference Collection Catalogue, The Language Studies Unit, University of Aston in Birmingham.

Faulkner, A. (1985) Nursing: A Creative Approach, London: Bailière Tindall.

Fincham, M. (1982) 'Hospital communication', in A. Waters (ed.), Issues in ESP, (Lancaster Practical Papers in English Language Education, Vol. 5), Oxford: Pergamon, pp. 63-98.

Fink, A. and J. Kosecoff (1985) How to Conduct Surveys: A Step-by-Step Guide, Beverly Hills & London: Sage.

Forrest, J. and M. Watson (1981) Practical Nursing and Anatomy for Pupil Nurses, London: Edward Arnold.

Fowler, W. (1987) The Right Word, Walton-on-Thames: Nelson.

Freeborn, D., D. Langford and P. French (1986) Varieties of English: An Introduction to the Study of Language, London: Macmillan Education.

Gimson, A. (1981) 'Pronunciation in EFL dictionaries', Applied Linguistics, Vol. 2, No. 3, pp. 250-262.

Godman, A. (1975) 'The language of science from the viewpoint of the writer of science books', paper presented at 10th regional seminar, (symposium 6), SEAMEO Regional English Language Centre, Singapore.

Godman, A. and E. Payne (1981)) 'A taxonomic approach to the lexis of science', in L. Selinker, E. Tarone and V. Hanzeli (eds.), pp. 23-39.

Greenbaum, S., C. Meyer and J. Taylor (1984) 'The image of the dictionary for American college students', Dictionaries, No. 6, Indiana University, pp. 31-52.

Griffin, P. (1985) Dictionaries in the ESL Classroom, M.A. dissertation, Southern Illinois University.

Haas, M. (1962) 'What belongs in the bilingual dictionary', in F. Householder and S. Saporta (eds.) pp. 45-50.

Halliday, M. (1966) 'Lexis as a linguistic level', in C. Bazell, J. Catford, M. Halliday and R. Robins (eds.), In Memory of J.R. Firth, London: Longman, pp. 148-162.

Halliday, M. (1973) Explorations in the Functions of Language, London: Edward Arnold.

Harrison, R. (1984) Textbook of Medicine with Relevant Physiology and Anatomy, London: Hodder and Stoughton.

Harrison, W., C. Prator and G. Tucker (1975) English Language Policy Survey of Jordan: a case study in language planning, Arlington, Virginia: Center for Applied Linguistics.

Hartmann, R. (1979a) Dictionaries and their Users, Exeter Linguistic Studies, Vol. 4, University of Exeter.

Hartmann, R. (1979b) 'Introduction: who needs dictionaries?', in R. Hartmann (ed.) (1979a), pp. 1-8.

Hartmann, R. (1981a) 'Dictionaries, learners, users: some issues in lexicography, a comparative review of four English language dictionaries', Applied Linguistics, Vol. 2, No. 3, pp. 297-303.

Hartmann, R. (1981b) 'Style values: linguistic approaches and lexicographical practice', Applied Linguistics, Vol. 2, No. 3, pp. 263-273.

Hartmann, R. (ed.) (1983a) Lexicography: principles and practice, London: Academic Press.

Hartmann, R. (1983b) 'The bilingual learner's dictionary and its uses', Multilingua, Vol. 2, No. 4, pp. 195-201.

Hartmann, R. (1983c) 'Theory and practice in dictionary-making', in R. Hartmann (ed.) (1983a), pp. 3-11.

Hartmann, R. (1983d) 'How to label contexts and varieties of usage', in R. Hartmann (ed.) (1983a), pp. 109-119.

Hartmann, R. (ed.) (1984) LEXeter '83 Proceedings, Tübingen: Niemeyer.

Hartmann, R. (1985) 'Contrastive text analysis and the search for equivalence in the bilingual dictionary', in K. Hyldgaard-Jensen and A. Zettersten (eds.), pp. 121-132.

Hartmann, R. (1986a) 'The dictionary in vocabulary learning, with particular reference to English learners of German', Mimeo.

Hartmann, R. (ed.) (1986b) The History of Lexicography, Vol. 40, Amsterdam and Philadelphia: John Benjamins

Hartmann, R. (1987) 'Four perspectives on dictionary use: a critical review of research methods', in A. Cowie (ed.) (1987b), pp. 11-28.

Hartmann, R. (forthcoming) 'Bilingual dictionary reference skills - some research priorities, Memio, [for publication in C. Butler (ed.) Festschrift for Walter Grauberg], University of Nottingham.

Hatherall, G. (1984) 'Studying dictionary use: some findings and proposals', in R. Hartmann (ed.), pp. 183-189.

Hawkey, R. (1980) 'Syllabus design for specific purposes', in Projects in Materials Design, (ELT documents: Special), London: The British Council, pp. 82-133.

Heltai, P. (1988) 'Contrastive analysis of terminological systems and bilingual technical dictionaries', International Journal of Lexicography, Vol. 1, No. 1, pp. 32-40.

Herbert, A. (1965) The Structure of Technical English, London: Longman.

Herbst, T. and G. Stein (1987) 'Dictionary-using skills: a plea for a new orientation in language teaching', in A. Cowie (ed.) (1987b), pp. 115-127.

Higgins, J. (1967) 'Hard facts (notes on teaching English to science students)', in J. Swales (ed.) (1985b), pp. 28-36.

Hoenigswald, H. (1962) 'Lexicography and grammar', in F. Householder and S. Saporta (eds.), pp. 103-110.

Holden, S. (ed.) (1977) English for Specific Purposes, Oxford: Modern English Publications.

Houghton, D. (1980) 'A collaborative approach to the teaching of vocabulary for accounting students', Team Teaching in ESP, (ELT documents 106), London: The British Council, pp. 24-33.

Houghton, D. (1984) 'Overseas students writing essays in English: learning the rules of the game', in G. James (ed.), pp. 47-57.

Householder, F. and S. Saporta (eds.) (1962) Problems in Lexicography, Vol. 21, (3rd ed.), Bloomington: Indiana University, (1975 imp.).

Howard, K. and J. Sharp (1983) The Management of a Student Research Project, Aldershot: Gower.

Hutchinson, T. and A. Waters (1980) 'Communication in the technical classroom: 'You just shove this little chappie in here like that'', in Projects in Materials Design, (ELT documents: Special), London: The British Council, pp. 7-36.

Hutchinson, T. and A. Waters (1982) 'Creativity in ESP materials or "HELLO! I'M A BLOOD CELL"', in A. Waters (ed.), Issues in ESP, (Lancaster Practical Papers in English Language Education, Vol. 5), Oxford: Pergamon, pp. 99-122.

Hutchinson, T. and A. Waters (1984) 'How communicative is ESP?', ELT Journal, Vol. 38, No. 2, pp. 108-113.

Hutchinson, T. and A. Waters (1987) English for Specific Purposes: A learning-centred approach, Cambridge University Press.

Hyldgaard-Jensen, K. and A. Zettersten (eds.) (1985) Symposium on Lexicography. Tübingen: Niemeyer.

Hymes, D. (1972) 'On communicative competence'. in J. Pride and J. Holmes (eds.), Sociolinguistics, Harmondsworth: Penguin, pp. 269-285.

Ibrahim, M. (1978) 'Patterns in spelling errors', ELT Journal, Vol. 32, No. 3, pp. 207-212.

Ilson, R. (1983) 'Etymological information: can it help our students?' ELT Journal, Vol. 37, No. 1, pp. 76-81.

Ilson, R. (ed.) (1985) Dictionaries, Lexicography and Language Learning, (ELT documents: 120), Oxford: Pergamon.

Ilson, R. (1987) 'Illustrations in dictionaries', in A. Cowie (ed.) (1987b), pp. 193-212.

Inman, M. (1978) ' Lexical analysis of scientific and technical prose', in M. Trimble, L. Trimble and K. Drobnic (eds.), English for Specific Purposes: Science and Technology, Oregon State University, pp. 242-249.

Iqbal, Z. (1987) Aspects of the Learner's Dictionary with Special Reference to Advanced Pakistani Learners of English, Ph.D. thesis, University of Aston in Birmingham.

Ivir, V. (1988) 'Collocations in dictionaries: monolingual and bilingual', in T. Burton and J. Burton (eds.), pp. 43-50.

Jackson, H. (1985) 'Grammar in the dictionary', in R. Ilson (ed.), pp. 53-59.

Jacob, S., C. Francone and W. Lossow (1982) Structure and Function in Man, (5th ed.), Philadelphia, London: Saunders.

Jain, M. (1981) 'On meaning in the foreign learner's dictionary', Applied Linguistics, Vol. 1, No. 3, pp. 274-286.

James, G. (ed.) (1984) The ESP Classroom - Methodology, Materials, Expectations, Exeter Linguistic Studies, Vol. 7, University of Exeter.

Jordan, Government of. (1964) Law No. 16 of 1964. Culture and Education Law (English translation) mimeographed: Amman.

Jordan, Ministry of Education (1971) New Living English for Jordan, Pupils' Book Three, Amman: Ministry of Education and Longman, (imp. 1986).

Jordan, Ministry of Education (1971) English Curriculum: Secondary Stage, (1st ed.), (in English), Amman: Curricula and Textbooks Division, Ministry of Education.

Jordan, Ministry of Education (1974) Oxford Secondary English Course: Jordan, Book One, Amman: Ministry of Education, and Oxford University Press.

Jordan, Ministry of Education (1976) Oxford Secondary English Course: Jordan, Book Three, Amman: Ministry of Education, and Oxford University Press.

Jordan, Ministry of Education (1978) Oxford Secondary English Course: Jordan, Book Two, Amman: Ministry of Education, and Oxford University Press.

Jordan, Ministry of Education (1984) Compulsory Stage English Curriculum, (1st ed.), (in English), Amman: Curriculum Directorate, Ministry of Education.

Jordan, Ministry of Education (1987a) PETRA: Progress in English through Relevant Activities for Jordan, Workbook 3, (trial ed.), Amman: Curriculum Directorate, Ministry of Education and Longman.

Jordan, Ministry of Education (1987b) PETRA: Progress in English through Relevant Activities for Jordan, Students' Book 3, (trial ed.), Amman: Curriculum Directorate, Ministry of Education and Longman.

Jordan, Ministry of Youth (1987) Al-Shabab, (in Arabic), No. 146, Amman: Ministry of Youth.

Jordan, R. and A. Mathews (1978) 'English for Academic Purposes: practical material for the listening comprehension and writing needs of overseas students', English for Specific Purposes, (ELT documents 101), London: The British Council, pp. 7-22.

Jordan, R.R. (1980) Academic Writing Course, London: Collins.

Kane, E. (1984) Doing Your Own Research, London and New York: Marion Boyars (1987 imp.).

Kennedy, C. and R. Bolitho (1984) English for Specific Purposes, London: Macmillan.

Kerr, R. and J. Smith (1978) Nucleus: Nursing Science, London: Longman.

Kharma, N. (1981) 'Analysis of Arab students' errors in using the English articles', 7th monograph in linguistics, Vol. 2, Annals of the College of Arts, Kuwait University, pp. 6-57.

Kipfer, B. (1984) Workbook on Lexicography: A Course for Dictionary Users with a Glossary of English Lexicographical Terms, Exeter Linguistic Studies, Vol. 8, University of Exeter.

Kipfer, B. (1985) Dictionaries and the Intermediate Student: Communicative Needs and the Development of User Reference Skills, M.Phil. thesis, University of Exeter.

Kipfer, B. (1987) 'Dictionaries and the intermediate student: communicative needs and the development of user reference skills', in Cowie (ed.) (1987b), pp.44-54.

Krashen, S. (1984) Writing: Research, Theory, and Applications, Oxford: Pergamon.

Krebs, W. (1988) 'Horses for courses: the design of small dictionaries for particular users', in T. Burton and J. Burton (eds.), pp. 51-62.

Kruse, A. (1979) 'Vocabulary in context', ELT Journal, Vol. 33, No. 3, pp. 207-213.

Leech, G. (1983) Principles of Pragmatics, London: Longman.

Leeuw, M. and E. Leeuw (1965) Read Better, Read Faster: A New Approach to Efficient Reading, Harmondsworth: Penguin (1984 imp.).

Leonard, P. (1983) Building a Medical Vocabulary, Philadelphia and London: Saunders.

Lewis, R. and M. Pugmire (1980) How to Use Your Dictionaries, Cambridge: National Extension College.

López, C. (1987) 'The dictionary: which, when, and how in advanced reading', English Teaching Forum, Vol. 25, No. 1, pp. 14-16.

Lyons, J. (ed.) (1970) New Horizons in Linguistics, Harmondsworth: Penguin.

MacFarquhar, P. and J. Richards (1983) 'On dictionaries and definitions', RELC Journal, Vol. 14, No. 1, pp.111-124.

Mackay, R. and A. Mountford (eds.) (1978) English for Specific Purposes, London: Longman.

Mackay, R., B. Barkman and R. Jordan (eds.) (1979) Reading in a Second Language: Hypotheses, Organization, and Practice, Rowley: Newbury House.

Magay, T. (1979) 'Problems of indicating pronunciation in bilingual dictionaries with English as the source language', in R. Hartmann (ed.) (1979a), pp. 98-103.

Magay, T. (1984) 'Technical or general: problems of vocabulary selection in a medium-size bilingual dictionary', in R. Hartmann (ed.), pp. 221-225.

Maher, J. (1986) 'The development of English as an international language of medicine', Applied Linguistics, Vol. 7, No. 2, pp. 206-219.

Maingay, S. and M. Rundell (1987) 'Anticipating learners' errors - implications for dictionary writers', in A. Cowie (ed.) (1987b), pp. 128-135.

Marckwardt, A. (1980) 'Getting the most out of the dictionary', in D. Bouchard and L. Sparenta, A TEFL Anthology: Selected Articles from the English Teaching Forum 1973-78, Washington: English Teaching Division, International Communication Agency, pp. 160-166.

Martin, W. (1985) 'Reflections on learner's dictionaries', in K. Hyldgaard-Jensen and A. Zettersten (eds.), pp. 169-181.

Mathews, S. (1987) 'Integrated Training Course Design: The University of Jordan In-service Diploma in TEFL', paper given at the 21st International Conference of IATEFL in Belgium.

Maxwell, A. (1978) Basic Statistics for Medical and Social Science Students, London and New York: Chapman and Hall (1983 imp.).

McArthur, T. (1986a) Worlds of Reference, Cambridge University Press.

McArthur, T. (1986b) 'Thematic lexicography', in R. Hartmann (ed.) (1986b), pp. 157-166.

McDonough, J. (1984) ESP in Perspective: A Practical Guide, London: Collins.

Meara, P. (1980) 'Vocabulary acquisition: a neglected aspect of language learning', Language Teaching and Linguistics: Abstracts, Vol. 13, No. 4, pp. 221-246.

Michiels, A. and J. Noël (1984) 'The pro's and con's of a controlled defining vocabulary in a learner's dictionary', in R. Hartmann (ed.), pp. 385-394.

Moulin, A. (1979) 'Dictionaries - general, technical, specialist, etc. - and how they often leave (foreign) advanced learners in the lurch', in R. Hartmann (ed.) (1979a), pp. 76-81.

Moulin, A. (1983) 'LSP dictionaries for EFL learners', in R. Hartmann (ed.) (1983a), pp. 144-152.

Moulin, A. (1987) 'The dictionary and encoding tasks', in A. Cowie (ed.) (1987b), pp. 105-114.

Mukattash, L. (1978) 'Further Studies in Jordanian English', mimeo, Department of English, University of Jordan.

Mukattash, L. (1980a) 'Yes/no questions and the contrastive analysis hypothesis, ELT Journal, Vol. 34, No. 2, pp. 133-145.

Mukattash, L. (1980b) 'English language proficiency in Jordan: educational, socio-economic and other factors', DIRASAT, (University of Jordan), Vol. 7, No. 2, pp. 7-31.

Mukattash, L. (1981a) 'Problems in error analysis', Papers and Studies in Contrastive Linguistics, Vol. 13, pp. 261-273.

Mukattash, L. (1981b) 'The evaluation and testing of English in Jordan: a critique', ELT Journal, Vol. 35, No. 4, pp. 450-454.

Mukattash, L. (1981c) 'Wh. questions in English: a problem for Arab learners', IRAL, Vol. 19, No. 4, pp. 317-332.

Munby, J. (1978) Communicative Syllabus Design, Cambridge University Press.

Nattinger, J. (1988) 'Some current trends in vocabulary teaching', in R. Carter and M. McCarthy (eds.), pp. 62-82.

Nelsen, D. (1976) 'Contrastive semantics in vocabulary instruction', TESOL Quarterly, Vol. 10, No. 1, pp. 99-103.

Nesi, H. (1984) Dealing with Lexical Errors, M.Sc. dissertation, University of Aston in Birmingham.

Nesi, H. (1987) 'Do dictionaries help students write?', in T. Bloor and J. Norrish (eds.), Written Language, (British Studies in Applied Linguistics 2), London: Centre for Information on Language Teaching and Research.

Nuttall, C. (1982) Teaching Reading Skills in a Foreign Language, London: Heinemann Educational.

O'Brien, T. and R. Jordan (1985) Developing Reference Skills, London: Collins.

Oláh, E. (1984) 'How special is special English?', in A. Pugh and J. Ulijn (eds.), pp. 223-226

Opitz, K. (1979) 'Technical dictionaries: testing the requirements of the professional user', in R. Hartmann (ed.) (1979a), pp. 89-95.

Opitz, K. (1983a) 'Special-purpose lexicography: dictionaries for technical use', in R. Hartmann (ed.) (1983a), pp. 163-180.

Opitz, K. (1983b) 'The segmental dictionary', in R. Hartmann (ed.) (1983a), pp. 53-64.

Országh, L. (1969) 'Wanted: better English dictionaries', ELT Journal, Vol. 23, No. 3, pp.216-222.

Pearce, E. (1980) A General Textbook of Nursing, London: Faber and Faber.

Phillips, M. (1983a) Lexical Macrostructure in Science Text, Ph.D. thesis, University of Birmingham.

Phillips, M. (1983b) 'Technical terminology and the creation of lexical meaning', paper presented at the 6th annual collouium of GERAS, Paris.

Picht, H. and J. Draskau (1985) Terminology: An Introduction, Department of Linguistic and International Studies Monographs, University of Surrey.

Piotrowski, T. (1987) 'Indication of English pronunciation in bilingual dictionaries', Applied Linguistics, Vol. 8, No. 1, pp. 39-47.

Prodromou, L. (1985) Medicine: Developing eading Skills in English, Oxford: Pergamon.

Pugh, A. and J. Ulijn (eds.) (1984) Reading for Professional Purposes: Studies and Practices in Native and Foreign Languages, London: Heinemann Educational.

Quinn, F. (1980) The Principles and Practice of Nurse Education, London: Croom Helm.

Quirk, R. (1973) 'The social impact of dictionaries in the UK', in R. McDavid and A.R. Duckert (eds.), Lexicography in English, (Annals of the New York Academy of Sciences), Vol. 211, No. 8, pp. 76-88.

Rauf, S. (1987) 'Developing reference skills', English Teaching Forum, Vol. 25, No. 1, pp. 17-19.

Reif, J. (1987) 'The development of a dictionary concept: an English learner's dictionary and an exotic alphabet', in A. Cowie (ed.) (1987b), pp. 146-158.

Richards, J. (1976) 'The role of vocabulary teaching' TESOL Quarterly, Vol. 10, No, 1, pp. 77-89.

Rixon, S. (1981) 'The design of materials to foster particular listening strategies', The Teaching of Listening Comprehension, (ELT documents: Special), London: The British Council, pp. 68-106.

Roberts, R. (1984) 'Contextual dictionaries for Languages for Special Purposes', in A. Pugh and J. Ulijn (eds.), pp. 154-165.

Robinson, P. (1980) ESP (English for Specific Purposes), Oxford: Pergamon.

Robinson, P. (1983) 'ESP, communicative language teaching, and the future', in K. Johnson and D. Porter (eds.), Perspectives in Communicative Language Teaching, London: Academic Press, pp. 159-170.

Roper, N. (1982) Principles of Nursing, London: Churchill Livingstone.

Rossner, R. (1985) 'The learner as lexicographer: using dictionaries in second language learning', in R. Ilson (ed.), pp. 95-102.

Rowntree, D. (1981) Statistics Without Tears: A Primer for Non-mathematicians, Harmondsworth: Penguin (1987 imp.).

Rudska, B., J. Channell, Y. Putseys and P. Ostyn (1981) The Words You Need, London: Macmillan.

Ryan, M. (1984) 'An approach to the teaching of word roots to paramedical students', EMP Newsletter, (Kuwait University), Vol. 1, No. 2, pp. 19-23.

Sager, J. (1984) 'Terminology and the technical dictionary', in R. Hartmann (ed.), pp. 315-326.

Sager, J., D. Dungworth and P. McDonald (eds.) (1980) English Special Languages: Principles and Practice in Science and Technology, Wiesbaden: Brandstetter Verlag.

Salager, F. (1985) 'Specialist medical English lexis: classificatory framework and rhetorical function - a statistical approach', EMP Newsletter, (Kuwait University), Vol. 2, No. 2, pp. 5-17.

Saville-Troike, M. (1982) The Ethnography of Communication: An Introduction, Oxford: Blackwell.

Scholfield, P. (1980) 'Evaluating selection policy and grammatical and semantic information in an EST dictionary', Fachsprache (Vienna), Vol. 2, No. 3, pp. 98-109.

Scholfield, P. (1982a) 'Using the English dictionary for comprehension', TESOL Quarterly, Vol. 16, No. 2, pp. 185-194.

Scholfield, P. (1982b) 'The role of bilingual dictionaries in ESL/EFL: a positive view', Guidelines, Vol. 4, No. 1, pp. 84-98.

Seidl, J. and W. McMordie (1978) English Idioms and How to Use Them, Oxford University Press.

Selinker, L., E. Tarone and V. Hanzeli (eds.) (1981) English for Academic and Technical Purposes: Studies in Honour of Louis Trimble, Rowley and London: Newbury House.

Sherman, D. (1978) 'Special purpose dictionaries', Cahiers de Lexicologie, Vol. 32, No. 2, pp. 82-102.

Sim, D. and B. Laufer-Dvorkin (1984) Vocabulary Development, London: Collins.

Sinclair, J. (1984) 'Lexicography as an academic subject', in R. Hartmann (ed.), pp. 3-12.

Sinclair, J. (1985) 'Lexicographic evidence', in R. Ilson (ed.), pp. 81-94.

Sinclair, J. (1987) The Dictionary of the Future, Collins English Dictionary Annual Lecture, given at the University of Strathclyde, 6 May, 1987.

Sinclair, J. and A. Renouf (1988) 'A lexical syllabus for language learning', in R. Carter and M. McCarthy (eds.), pp. 140-160.

Skehan, P. (1980) 'Team-teaching and the role of the ESP teacher', Study Modes and Academic Development of Overseas Students, (ELT documents: 109), London: The British Council, pp. 23-37.

Smith, A. (1982) 'The importance of attitude in foreign-language learning', The Art of TESOL: Selected Articles from the English Teaching Forum, Part 1, special edition, Washington: English Teaching Division, International Communication Agency, pp. 20-25.

Snell-Hornby, M. (1984) 'The bilingual dictionary - help or hindrance?', in R. Hartmann (ed.), pp. 274-281.

Snell-Hornby, M. (1987) 'Towards a learner's bilingual dictionary', in A. Cowie (ed.) (1987b), pp. 159-170.

Spatola, A. (1982) Mastering Medical Language, Englewood Cliffs: Prentice-Hall.

Stein, G. (1984) 'Towards a theory of lexicography: principles and/vs. practice in modern English dictionaries', in R. Hartmann (ed.), pp. 124-130.

Strevens, P. (ed.) (1978) In Honour of A.S. Hornby, Oxford University Press.

Strevens, P. (1980) Teaching English as an International Language, from Practice to Principle, Oxford: Pergamon.

Summers, D. (1988a) 'English language teaching dictionaries: past, present, and future', English Today, Vol. 14, No. 2, pp. 10-16.

Summers, D. (1988b) 'The role of dictionaries in language learning', in R. Carter and M. McCarthy (eds.), pp. 111-125.

Swales, J. (1977) 'ESP in the Middle East', in S. Holden (ed.), pp. 36-38.

Swales, J. (1980a) 'The educational environment and its relevance to ESP programme design', in Projects in Materials Design, (ELT documents: Special), London: The British Council, pp. 61-70.

Swales, J. (1980b) 'ESP - the textbook problem', ESP Journal, Vol. 1, No. 1, pp. 11-23.

Swales, J. (1983) 'Vocabulary work in LSP - a case of neglect?', in F. Redard and S. Wyler (eds.), Langues de Spécialité', Special Number of CILA Bulletin, (Neuchâtel), pp. 21-33.

Swales, J. (1984a) 'A review of ESP in the Arab World 1977-1983 - trends, developments and retrenchments', in J. Swales and H. Mustafa (eds.), pp. 9-21.

Swales, J. (1984b) 'Thoughts on, in and outside the classroom', in G. James (ed.), pp. 7-16.

Swales, J. (1985a) 'ESP - the heart of the matter or the end of the affair?', in R. Quirk and H. Widdowson (eds.), English the World - Teaching and Learning the Language and Literatures, Cambridge University Press, pp. 212-227.

Swales, J. (ed.) (1985b) Episodes in ESP, Oxford: Pergamon.

Swales, J. and H. Mustafa (eds.) (1984) English for Specific Purposes in the Arab World, The Language Studies Unit, University of Aston in Birmingham.

Tadros, A. (1979) 'Arabic interference in the written English of Sudanese students - relativisation', ELT Journal, Vol. 33, No. 3, pp. 234-239.

Tawfiq, A. (1984) 'Designing a communicative EST reading skills course for students of the faculty of medicine in Iraq: a case study', in J. Swales and H. Mustafa (eds.), pp. 85-98.

Teichroew, F. (1982) ' Receptive versus productive vocabulary: a survey', Interlanguage Studies Bulletin, Vol. 6, No. 2, pp. 5-33.

Temple, M. (1985) A Pocket Guide to Spelling, London: John Murray.

Thompson, G. (1987) 'Using bilingual dictionaries', ELT Journal, Vol. 14, No. 4, pp. 282-286.

Tomaszczyk, J. (1979) 'Dictionaries: users and uses', Glottodidactica, Vol. 12, pp. 103-119.

Tomaszczyk, J. (1981) 'Issues and developments in bilingual pedagogical lexicography', Applied Linguistics, Vol. 2, No. 3, pp. 287-296.

Tomaszczyk, J. (1983) 'The case for bilingual dictionaries for foreign language learners', in R. Hartmann (ed.) (1983a), pp. 41-51.

Tomaszczyk, J. (1987) 'FL learners' communication failure: implications for pedagogical lexicography', in A. Cowie (ed.) (1987b), pp. 137-145.

Tono, Y. (1984) On the Dictionary User's Reference Skills, B.Ed. dissertation, Tokyo Gakugie University.

Trimble, L. (1985) English for Science and Technology: A Discourse Approach, Cambridge University Press.

Underhill, A. (1980) Use Your Dictionary: A Practice Book for Users of OXFORD ADVANCED LEARNER'S DICTIONARY OF CURRENT ENGLISH and OXFORD STUDENT'S DICTIONARY OF CURRENT ENGLISH, Oxford University Press, (1985 imp.).

Vanparreren, C. and M. Schouten-Vanparreren (1982) 'Contextual guessing: a trainable reader strategy', System, Vol. 9, No. 3, pp. 235-241.

Wallace, M. (1982) Teaching Vocabulary, London: Heinemann Educational.

Wells, J. (1985) 'English pronunciation and its dictionary representation', in R. Ilson (ed.), pp. 45-51.

Whitcut, J. (1979) Learning with LDOCE, London: Longman, (1981 imp.).

Whitcut, J. (1988) 'Lexicography in simple language', International Journal of Lexicography, Vol. 1, No. 1, pp. 49-55.

Widdowson, H. (1979) Explorations in Applied Linguistics, Oxford University Press

Widdowson, H. (1981) 'English for specific purposes: criteria for course design', in L. Selinker, E, Tarone and V. Hanzeli (eds.), pp. 1-11.

Widdowson, H. (1983) Learning Purpose and Language Use, Oxford University Press.

Widdowson, H. (1984) Explorations in Applied Linguistics 2, Oxford University Press.

Wiegand, H. (1984) 'On the structure and contents of a general theory of lexicography', in R. Hartmann (ed.), pp. 13-30.

Wierzbicka, A. (1985) Lexicography and Conceptual Analysis, Ann Arbor: Karoma.

Wilkins, D. (1972) Linguistics in Language Teaching, London: Edward Arnold, (1980 imp.).

Wilkins, D. (1974) Second-Language Learning and Teaching, London: Edward Arnold, (1984 imp.).

Wilkins, D. (1976) Notional Syllabuses, Oxford University Press.

Willcott, P. (1978) 'Problems of definitiveness in the written English of Arabic speakers', ELT Journal, Vol. 33, No. 1, pp. 67-73.

Williams, R. (1981) 'Lexical familiarization in content area textbooks', in J. Chapman (ed.), The Reader and the Text, London: Heinemann Educational, pp. 49-59.

Woods, A., P. Fletcher and A. Hughes (1986) Statistics in Language Studies, Cambridge University Press.

LIST OF ABBREVIATIONS

BND	= BAILLIERE'S NURSES' DICTIONARY
CELD	= COLLINS ENGLISH DICTIONARY
COBUILD	= COLLINS Birmingham University International Database
CULD	= CHAMBERS UNIVERSAL LEARNERS' DICTIONARY
JD	= Jordanian Dinar
EFL	= English as a Foreign Language
ELT	= English Language Teaching
ESP	= English for Specific Purposes
EST	= English for Science and Technology
G.S.E.C.E.	= General Secondary Education Certificate Examination
GPE	= General Purpose English
LDOCE	= LONGMAN DICTIONARY OF CONTEMPORARY ENGLISH
LDOCU	= LONGMAN DICTIONARY OF SCIENTIFIC USAGE
LSP	= Language/s for Specific Purposes
MDOE	= MODERN DICTIONARY OF ELECTRONICS
OALD	= OXFORD ADVANCED LEARNER'S DICTIONARY OF CURRENT ENGLISH
TEFL	= Teaching of English as a Foreign Language
TESOL	= Teaching of English as a Second Language
TOEFL	= Test of English as a Foreign Language
UNRWA	= United Nations Relief and Works Agency
UPE	= University Proficiency Examination
WNWD	= WEBSTER'S NEW WORLD DICTIONARY OF THE AMERICAN LANGUAGE

APPENDIX 1: THE STUDENTS' QUESTIONNAIRE

(translation from Arabic)

The Language Centre

The University of Jordan

Dear Students,

This questionnaire is part of a PhD research project I am currently carrying out at the University of Exeter in the U.K. The research is concerned with the use of dictionaries among nursing students at Jordanian Universities in relation to the problems of learning English.

The information you will give by answering the questions in this questionnaire is especially important as it will contribute to the development of the specific-purpose English language courses that are currently offered in the Faculties of Nursing in cooperation with the Language Centres.

While appreciating your participation in this study, I would like to assure you that all information in this questionnaire will be treated strictly confidentially and will be used for research purposes only.

Your cooperation is greatly appreciated.

Turki Diab

The University of Exeter/U.K.

Section One

Please give the following information about yourself by putting a circle around the number that indicates your answer or by filling the blanks according to the nature of the question. In case you do not know the answer please write "I do not know".

1. male   2. female

2. level of study   1. first year   2. second year
  3. third year   4. fourth year

3. accumulative average: (our of 100%) __________

4. type of secondary education certificate:

   1. Jordanian

   2. other Arab countries (please specify) __________

   3. other foreign countries (please specify) __________

5. type of secondary school you graduated from:

   1. government

   2. private

   3. army

   4. other (please specify) __________

6. name of secondary school you graduated from:

   ____

7. general aggregate in G.S.E.C.E. (out of 100%) __________

8. score in English in G.S.E.C.E. (out of 100%) __________

9. name of district/governerate in which you finished secondary

   education ____________________

10. Have you been accepted to specialisations/faculties other than nursing when you finished the secondary education?

    1. yes   2. no

    If your answer is yes, please specify these specilisations or faculties.

Section Two

Please put a circle around the number that indicates your answer. The numbers mean:

1. never 2. rarely 3. sometimes 4. often 5. always

1. Before joining the University, have you used dictionaries (regardless of their type or size)?

in the Elementary Stage.	1 2 3 4 5
in the Preparatory Stage.	1 2 3 4 5
in the Secondary Stage.	1 2 3 4 5

2. In your University study, do you use the following dictionaries?

AL-MAWRID (large size) English-Arabic	1 2 3 4 5
AL-MAWRID AL-WASIT (medium) English Arabic	1 2 3 4 5
AL-MAWRID (pocket) English-Arabic	1 2 3 4 5
Elias' MODERN DICTIONARY (large size) English-Arabic	1 2 3 4 5
Elias' POCKET DICTIONARY English-Arabic	1 2 3 4 5
Elias MODERN DICTIONARY (large size) Arabic-English	1 2 3 4 5
HITTI'S MEDICAL DICTIONARY English-Arabic	1 2 3 4 5
DICTIONARY OF SCIENTIFIC TECHNICAL AND ENGINEERING TERMS	1 2 3 4 5
POCKET MEDICAL DICTIONARY (Dr Shihabi) English-Arabic	1 2 3 4 5
OXFORD ENGLISH-ARABIC DICTIONARY OF CURRENT USAGE	1 2 3 4 5
OXFORD ADVANCED LEARNER'S DICTIONARY	1 2 3 4 5
LONGMAN DICTIONARY OF CONTEMPORARY ENGLISH	1 2 3 4 5
NURSES' DICTIONARY	1 2 3 4 5
WEBSTER'S NEW COLLEGIATE DICTIONARY	1 2 3 4 5

1. never 2. rarely 3. sometimes 4. often 5. always

NEW WEBSTER'S DICTIONARY OF THE ENGLISH LANGUAGE	1 2 3 4 5
COLLINS ENGLISH DICTIONARY	1 2 3 4 5
Any other dictionaries not mentioned? (please specify)	
________________________	1 2 3 4 5
________________________	1 2 3 4 5

3. Do you use a dictionary in the following activities, and how often?

- understanding	
- assigned textbooks	1 2 3 4 5
- teachers handouts	1 2 3 4 5
- scientific journals	1 2 3 4 5
- what is said in lectures	1 2 3 4 5
- what is said in seminars	1 2 3 4 5
- what is said in hospital rounds	1 2 3 4 5
- writing	
- research essays	1 2 3 4 5
- technical reports	1 2 3 4 5
- nursing care plans	1 2 3 4 5
- preparing oral presentations	1 2 3 4 5
- any other activities? Please specify:	
________________________	1 2 3 4 5

1. never 2. rarely 3. sometimes 4. often 5. always

4. Do you use dictionaries to look up:

-meaning?	1 2 3 4 5
-pronunciation?	1 2 3 4 5
-spelling?	1 2 3 4 5
-word division into syllables e.g. (dic.tion.a.ry)?	1 2 3 4 5
-Latin or Greek origins (e.g. the word cardiograph comes from the Latin word kardia which means the heart and the word graphein which means to write?	1 2 3 4 5
-the appropriate preposition that can be used with certain words (e.g. deal with, divide into)?	1 2 3 4 5
-other grammatical information (e.g. whether a verb is transitive or intransitive, whether a noun is countable or uncountable)	1 2 3 4 5
-any other purposes? Please specify: ______________________	1 2 3 4 5

5. Do you ask the nursing teachers about:

-meanings of words/phrases?	1 2 3 4 5
-pronunciation of words/phrases?	1 2 3 4 5
-type of dictionary you should use?	1 2 3 4 5
-symbols used in dictionaries?	1 2 3 4 5
-how to use certain dictionaries?	1 2 3 4 5

6. Do you ask your fellow students about:

-meanings of words/phrases?	1 2 3 4 5
-pronunciation of words/phrases?	1 2 3 4 5
-type of dictionary you should use?	1 2 3 4 5
-symbols used in dictionaries?	1 2 3 4 5
-how to use certain dictionaries?	1 2 3 4 5

1. <u>never</u> 2. <u>rarely</u> 3. <u>sometimes</u> 4. <u>often</u> 5. <u>always</u>

7. Do you ask the <u>teachers of English</u>					
-meanings of words/phrases?	1	2	3	4	5
-pronunciation of words/phrases?	1	2	3	4	5
-type of dictionary you should use?	1	2	3	4	5
-symbols used in dictionaries?	1	2	3	4	5
-how to use certain dictionaries?	1	2	3	4	5
8. Do you use a dictionary in the classroom during					
-nursing classes	1	2	3	4	5
-English language classes	1	2	3	4	5
9. Do you use the dictionaries available in the University Library?	1	2	3	4	5
10. Do you use more than one dictionary to look up the same word or phrase?	1	2	3	4	5
11. While reading, do you look up the meanings of all unfamiliar words or phrases?	1	2	3	4	5
12. While reading, do you try to guess the meanings of unfamiliar words or phrases?	1	2	3	4	5
13. Do you write the meanings of looked-up words on the text you read?	1	2	3	4	5
14. Do you write the meanings of new words/phrases in a special notebook?	1	2	3	4	5
15. Do you refer to the introduction of the dictionary pages?	1	2	3	4	5
16. Do you browse through the dictionary pages without having a particular purpose?	1	2	3	4	5

1. never 2. rarely 3. sometimes 4. often 5. always

17. Do you refer to the glossaries usually found at the beginning or the end of assigned textbooks? 1 2 3 4 5

18. Do you write the pronunciation of English words in Arabic letters in order to remember how to pronounce them (e.g. the word antibiotics is written as انتيبيوتكس)? 1 2 3 4 5

19. Do you refer to the illustrations (e.g. plates/ diagrams) found in dictionaries? 1 2 3 4 5

20. Do you refer to the general information in dictionary appendices (e.g. tables of measurements etc.)? 1 2 3 4 5

21. Do the nursing teachers help you in the following areas of the English language?

- grammar? 1 2 3 4 5
- pronunciation? 1 2 3 4 5
- meaning? 1 2 3 4 5
- spelling? 1 2 3 4 5

22. Do the English language teachers help you in the following areas of the English language:

- grammar? 1 2 3 4 5
- pronunciation? 1 2 3 4 5
- meaning? 1 2 3 4 5
- spelling? 1 2 3 4 5

Section Three

Please put a circle around the number that indicates your opinion next to each of the following statements below. The numbers mean:

1. strongly agree 2. agree 3. disagree 4. strongly disagree 5. I don't know

23. What do you think of the following statements?

-It is important for nursing students to use dictionaries in their university study.	1	2	3	4	5
-All dictionaries contain the same information	1	2	3	4	5
-In general, dictionary using is boring.	1	2	3	4	5
-Nursing students need to refer to a dictionary when they write in English.	1	2	3	4	5
-To depend on one dictionary is not enough.	1	2	3	4	5
-Nursing students can generally understand the content of their assigned textbooks without reference to a dictionary.	1	2	3	4	5
Nursing students should be allowed to use dictionaries during:					
-nursing examinations	1	2	3	4	5
-English language examinations	1	2	3	4	5
-Every nursing student should have at least one dictionary.	1	2	3	4	5
-It is better for nursing students to learn English through nursing texts/subjects.	1	2	3	4	5
-The present English language courses run by the Language Centre provide sufficient information on dictionaries and how to use them.	1	2	3	4	5
-Nursing students should avoid using pocket dictionaries.	1	2	3	4	5
-It is important for nursing students to use a medical/nursing dictionary in their study.	1	2	3	4	5

1. strongly agree 2. agree 3. disagree 4. strongly disagree
5. I don't know

-English-English dictionaries are more useful to nursing students than English-Arabic dictionaries.	1 2 3 4 5
-AL-MAWRID is the most useful dictionary for nursing students.	1 2 3 4 5
-In general, studying nursing is enjoyable.	1 2 3 4 5
-It would be better if the present English Language courses included exercises on English-English dictionaries and how to use them.	1 2 3 4 5

Section Four

Please put a circle around the number that indicates the amount of need for help you require in the English language through the English language courses offered by the Faculty of Nursing in cooperation with the Language Centre.

The numbers mean:

1. no need 2. little need 3. medium need 4. much need
5. great/immediate need

24. In which areas of the following do you need help with English:

A. Understanding spoken English used during:

- lectures	1 2 3 4 5
- seminars	1 2 3 4 5
- hospital rounds	1 2 3 4 5
- oral examinations	1 2 3 4 5

B. Understanding written English used in:

- assigned textbooks	1 2 3 4 5
- teacher handouts	1 2 3 4 5
- scientific journals	1 2 3 4 5
- written examinations	1 2 3 4 5

1. <u>no need</u> 2. <u>little need</u> 3. <u>medium need</u> 4. <u>much need</u>
5. <u>great/immediate need</u>

C. Writing:

- research essays	1	2	3	4	5
- technical reports	1	2	3	4	5
- nursing care plans	1	2	3	4	5
- answers in written examinations	1	2	3	4	5
- notes when listening to spoken English	1	2	3	4	5

D. Speaking

- during lectures	1	2	3	4	5
- during seminars	1	2	3	4	5
- during hospital rounds	1	2	3	4	5
- giving an oral presentation	1	2	3	4	5
- answering oral examination questions	1	2	3	4	5

25. How much help do you need in the following?

First -

- speaking about nursing topics	1	2	3	4	5
- writing about nursing topics	1	2	3	4	5
- understanding spoken texts about nursing topics	1	2	3	4	5
- understanding written texts about nursing topics	1	2	3	4	5

Second -

- grammar	1	2	3	4	5
- pronunciation	1	2	3	4	5
- vocabulary meaning	1	2	3	4	5
- common abbreviations in nursing	1	2	3	4	5
- spelling	1	2	3	4	5
- punctuation	1	2	3	4	5

1. <u>no need</u> 2. <u>little need</u> 3. <u>medium need</u> 4. <u>much need</u>
5. <u>great/immediate need</u>

- using English-English dictionaries	1 2 3 4 5
- any other areas? Please specify:	
- ______________________________	1 2 3 4 5
- ______________________________	1 2 3 4 5

26. Do you have any general comments you would like to give?
Please use the space below:

Thank you very much for your participation.

APPENDIX 2: THE NURSING STAFF QUESTIONNAIRE

The Language Centre

The University of Jordan

Dear colleague,

This questionnaire is part of a PhD research project I am currently carrying out at the University of Exeter in the U.K. The topic I am investigating is "Dictionary Use Among Jordanian University Students" in relation to the problems of learning English for academic purposes.

In any English for Specific Purposes programme, it is essential to integrate the views of the science-subjects staff into the English language teaching/learning process. Therefore, the information that you will give in this questionnaire will be useful in our joint efforts to improve the quality of the specific-purpose English courses that we currently offer to nursing students.

While being very grateful for your participation in this study, I would like to assure you that all information in this questionnaire will be treated strictly confidentially and used for research purposes only.

Your cooperation is greatly appreciated.

Turki Diab

The University of Exeter/ U.K.

Section One

Please fill in the blanks.

Q. 1. What is your nationality? ______________

Q. 2. How many year of university teaching experience do you have? ____

Q. 3. What is your latest academic qualification? ______________

Q. 4. Where did you study for your latest academic qualification? _________

Q. 5. What course/courses are you currently teaching?

a. ______________

b. ______________

c. ______________

Q. 6. What level/levels of students are you currently teaching?

(please put a / where appropriate)

a. first year ____________

b. second year ____________

c. third year ____________

d. fourth year ____________

Section Two

Please put a circle around the number that indicates the degree of importance you attach to the following. The numbers mean:

1. not important 2. slightly important 3. very important
4. extremely important 5. I do not know

Q. 7. Is it important for your students to use:

a - English-Arabic medical/nursing dictionaries?	1	2	3	4	5
b - English-Arabic non-medical/nursing dictionaries?	1	2	3	4	5
c - English-English medical/nursing dictionaries?	1	2	3	4	5
d - English-English non-medical/nursing dictionaries?	1	2	3	4	5
e - Arabic-English medical/nursing dictionaries?	1	2	3	4	5
f - Arabic-English non-medical/nursing dictionaries?	1	2	3	4	5

Q. 8. Is it important for your students to use dictionaries in the following:

a. understanding assigned textbooks?	1	2	3	4	5
b. understanding handouts?	1	2	3	4	5
c. understanding scientific journals?	1	2	3	4	5
d. understanding what is said in seminars?	1	2	3	4	5
e. understanding what is said in lectures?	1	2	3	4	5
f. understanding what is said in hospital rounds?	1	2	3	4	5
g. writing research essays?	1	2	3	4	5
h. writing technical reports?	1	2	3	4	5
i. writing nursing care plans?	1	2	3	4	5
j. preparing oral presentations?	1	2	3	4	5
k. other? please specify ________________________	1	2	3	4	5

1. <u>not important</u> 2. <u>slightly important</u> 3. <u>very important</u>
4. <u>extremely important</u> 5. <u>I do not know</u>

Q. 9. Is it important for your students to use their dictionaries for looking up:

a. meaning of words?	1 2 3 4 5
b. pronunciation of words?	1 2 3 4 5
c. Latin/Greek origins of words?	1 2 3 4 5
d. spelling of words?	1 2 3 4 5
e. appropriate prepositions that can be used with certain words (e.g. consist <u>of</u>)?	1 2 3 4 5
f. how words can be divided into syllables (e.g. dic.tion.a.ry)?	1 2 3 4 5
g. how words can be used grammatically correctly (e.g. passive/transitive/singular/countable etc.)?	1 2 3 4 5
h. other? Please specify: ____________________	1 2 3 4 5

<u>Section Three</u>

Please put a <u>circle</u> around the number that indicates the <u>frequency</u> in the following. The numbers mean:

1. <u>never</u> 2. <u>rarely</u> 3. <u>sometimes</u> 4. <u>often</u> 5. <u>always</u>

Q. 10. In general, do you use <u>English</u> in your teaching?	1 2 3 4 5
Q. 11. Do you explain meanings of words when your are teaching?	1 2 3 4 5
Q. 12. Do you teach your students how to pronounce words?	1 2 3 4 5
Q. 13. Do you use a dictionary when you prepare your lessons?	1 2 3 4 5
Q. 14. Do you encourage your students to buy certain dictionaries?	1 2 3 4 5
Q. 15. Do you advise your students on what dictionary/ies they should buy?	1 2 3 4 5
Q. 16. Do you explain to your students how to use dictionaries?	1 2 3 4 5

1. never 2. rarely 3. sometimes 4. often 5. always

Q. 17. In your examinations, do you use the following types of questions?

a. yes/no (or true/false)	1 2 3 4 5
b. essay	1 2 3 4 5
c. matching	1 2 3 4 5
d. defining	1 2 3 4 5
e. multiple choice	1 2 3 4 5
f. labelling	1 2 3 4 5
g. other? Please specify: ______________________	1 2 3 4 5

Section Four

Please put a circle around the number that indicates your opinion next to each of the statements below. The numbers mean:

1. strongly agree 2. agree 3. disagree 4. strongly disagree 5. I do not know

Q. 18. What is your opinion about the following statements:

- All dictionaries are the same	1 2 3 4 5
- In general, dictionary using is boring.	1 2 3 4 5
- Students should depend on one dictionary.	1 2 3 4 5
- Students should be allowed to use dictionaries in nursing examinations.	1 2 3 4 5
- Students should avoid using pocket dictionaries.	1 2 3 4 5
- Al-Mawrid (large size) is the dictionary that students should use.	1 2 3 4 5
- Students should be taught how to use dictionaries.	1 2 3 4 5
- English should be taught to students through nursing-related materials.	1 2 3 4 5
- Every student should have at least one dictionary.	1 2 3 4 5
- Students should know meanings of words in English.	1 2 3 4 5

1. strongly agree 2. agree 3. disagree 4. strongly disagree
5. I do not know

- Nursing subjects should be taught all in Arabic.	1 2 3 4 5
- Writing skills are less important for students than reading skills.	1 2 3 4 5
- In general, students enjoy studying nursing.	1 2 3 4 5
- Teaching medical/nursing terminology is the responsibility of the nursing staff	1 2 3 4 5
- Students should be encouraged to use English-English dictionaries.	1 2 3 4 5
- Nursing staff do not have the time to concentrate on students' language problems	1 2 3 4 5

Section Five

Please put a circle around the number that indicates the amount of need for help in the following. The numbers mean:

1. no need 2. little need 3. average need 4. much need
5. great need

Q. 19. Do your students need to use English in the following areas?

- Speaking in English:

- during lectures	1 2 3 4 5
- during seminars	1 2 3 4 5
- during hospital rounds	1 2 3 4 5
- during oral examinations	1 2 3 4 5
- giving an oral presentation	1 2 3 4 5
- other? Please specif:	
______________________________	1 2 3 4 5

1. <u>no need</u> 2. <u>little need</u> 3. <u>average need</u> 4. <u>much need</u> 5. <u>great need</u>

- <u>Writing in English</u>	
- answers to written examinations	1 2 3 4 5
- research essays	1 2 3 4 5
- technical reports	1 2 3 4 5
- notes when listening to spoken English	1 2 3 4 5
- nursing care plans	1 2 3 4 5
- other? Please specify:	
______________________________	1 2 3 4 5
- <u>Reading in English</u>	
- assigned textbooks	1 2 3 4 5
- handouts	1 2 3 4 5
- journals	1 2 3 4 5
- other? Please specify:	
______________________________	1 2 3 4 5
- <u>Understanding spoken English</u>	
a. - lectures	1 2 3 4 5
b. - seminars	1 2 3 4 5
c. - hospital rounds	1 2 3 4 5
d. - other? Please specify:	
______________________________	1 2 3 4 5

Section Six

Please put a circle around the number that indicates the amount of help your students require in the following. The numbers mean:

1. no help 2. little help 3. average help 4. much help
5. great help

How much help do your students require from the language centre in the English courses in the following:

1. no need 2. low need 3. average need 4. high need
5. I don't know

- understanding English grammar	1 2 3 4 5
- understanding meaning of general (non-medical/ nursing vocabulary	1 2 3 4 5
- understanding meaning of medical/nursing terminology	1 2 3 4 5
- pronunciation of general (non-medical/nursing vocabulary	1 2 3 4 5
- pronunciation of medical/nursing terminology	1 2 3 4 5
- using grammatically correct sentences in speaking	1 2 3 4 5
- using grammatically correct sentences in writing	1 2 3 4 5
- using correct spelling	1 2 3 4 5
- using correct punctuation	1 2 3 4 5
- using common medical/nursing abbreviations	1 2 3 4 5
- using English-English dictionaries	1 2 3 4 5
- other? Please specify: ______________________________	1 2 3 4 5

Section Seven

Q. 21. Please give the names of the dictionaries that you recommend for your students to use.

A. English-Arabic

1. ____________________

2. ____________________

3. ____________________

B. English-English

1. ____________________

2. ____________________

3. ____________________

Q. 22. Please use the space below to add any comments or suggestions you think appropriate concerning content of this questionnaire:

Your participation is greatly appreciated

APPENDIX 3: THE STUDENTS' FOLLOW-UP INTERVIEW

(implemented in Arabic)

The Language Centre
University of Jordan

Students' Interview

Introduction

1. A brief idea about the purpose of the study and in particular the nature of the interview.
2. Assuring the student about the confidentiality of information provided.
3. Guaranteeing the student's the right not to answer any question or part of question during the interview.

Section One: Background Information

1. Sex: 1. male 2. female

2. Level of study: 1. first year 2. second year
3. third year 4. fourth year

3. Cumulative Aggregate (if any): ________________

4. Type of secondary schooling:

1. Jordanian 2. non-Jordanian (specify)
____

3. Government 4. non-Government (specify)
____

5. G.S.E.C.E.:

   1. General Aggregate: ______________
   2. Score in the English examination: ______________
   3. District: ______________

Section Two: Pre-University Dictionary Use

1. Did you use dictionaries before joining the University?

   1. yes   2. no

   (If the answer is no move to question 8)

2. At which Stage/Class did you start using a dictionary?

   1. Stage: a. Elementary   b. Preparatory   c. Secondary

   2. Class: 5th, 6th, 7th, 8th, 9th, 10th, 11th, 12th

3. How did you start dictionary use?

	yes	no
1. English language teacher asked:	____	____
2. English language teacher advised:	____	____
3. Others advised:	____	____
4. Personal initiative:	____	____

   5. or? (specify) ____________________

4. What type/size of dictionary/ies were you using?

	yes	no
1. pocket English-Arabic	____	____
2. pocket English-English	____	____
3. pocket Arabic-English	____	____
4. large English-Arabic	____	____
5. large English-Arabic	____	____
6. large Arabic-English	____	____

5. What names of dictionaries were you using?

   1. ____________________

   2. ____________________

   3. ____________________

   4. ____________________

6. For what purposes were you using dictionaries?

	yes	no
1. meaning	____	____
2. pronunciation	____	____
3. spelling	____	____
4. grammar	____	____
5. other? (specify)	____	____

7. In what situations did you use dictionaries?

	yes	no	specifications
1. reading	____	____	______________
2. writing	____	____	______________
3. listening	____	____	______________
4. speaking	____	____	______________

8. Have you received any training in dictionary use?

8.1 1. yes 2. no (if the answer is no move to question 9)

8.2 Do you remember the Stage/Class when you first received such training?

Stage: 1. Elementary 2. Preparatory 3. Secondary

Class: 5th, 6th, 7th, 8th, 9th, 10th, 11th, 12th

8.3 Do you remember what type/s/size/s of dictionaries were used in such training?

1. Pocket English-Arabic
2. Pocket English-English
3. Large English-Arabic
4. Large English-English

8.4 What do you think of the training you have received?

1. useless 2. useful 3. very useful

1. not adequate 2. adequate

9. At which Stage of schooling have you used your dictionary most often?

   1. Elementary   2. Preparatory   3. Secondary

<u>Section Three</u>: Dictionary Use at University Level

	yes	no
10. Do you use a dictionary in your study?	----	----
11. How did you start dictionary use?		
1. a nursing teacher asked?	----	----
2. an ESP teacher asked?	----	----
3. a nursing teacher advised	----	----
4. an ESP teacher advised	----	----
5. others advised?	----	----
6. personal initiative?	----	----
12. What type.size of dictionary do you use?		
1. pocket English-Arabic	----	----
2. pocket English-English	----	----
3. pocket Arabic-English	----	----
4. large English-Arabic	----	----
5. large English-English	----	----
6. large Arabic-English	----	----

13. Would you mention the name/s of the dictionary/ies you are using? (names of dictionaries to be suggested as examples.)

   1. ______________________________

   2. ______________________________

   3. ______________________________

   4. ______________________________

   5. ______________________________

14. What do you use dictionaries for?

	yes	no	specifications
1. meaning	____	____	______________
2. pronunciation	____	____	______________
3. spelling	____	____	______________
4. syllabification	____	____	______________
5. etymology	____	____	______________
6. grammar	____	____	______________
7. illustrative materials	____	____	______________
8. information in appendices	____	____	______________

9. other? (specify) ___________________________

15. Do you use your dictionary in the following situations?

	yes	no	specifications
1. reading textbooks	____	____	______________
2. reading journals	____	____	______________
3. reading handouts	____	____	______________
4. understanding lectures	____	____	______________
5. understanding seminars	____	____	______________
6. understanding clinical rounds	____	____	______________
7. writing essays	____	____	______________
8. writing reports	____	____	______________
9. writing care plans	____	____	______________
10. speaking in lectures	____	____	______________
11. speaking in seminars	____	____	______________
12. speaking in clinical rounds	____	____	______________
13. preparing oral presentations	____	____	______________

16. In which courses do you use dictionaries most often? e.g. anatomy, medical-surgical-nursing.

1. ______________
2. ______________
3. ______________
4. ______________

Section Four: Strategies

17. Do you ask your nursing teachers about:

	yes	no	frequency
1. meanings of words	____	____	____________
2. pronunciation of words	____	____	____________
3. spelling	____	____	____________
4. how to use words grammatically	____	____	____________
5. dictionary use	____	____	____________
(specify)	________________________		

18. Do you ask your ESP teachers about:

	yes	no	frequency
1. meanings of words	____	____	____________
2. pronunciation of words	____	____	____________
3. spelling	____	____	____________
4. how to use words grammatically	____	____	____________
5. dictionary use	____	____	____________
(specify)	________________________		

19. Do you ask your fellow students about:

	yes	no	frequency
1. meanings of words	____	____	____________
2. pronunciation of words	____	____	____________
3. spelling	____	____	____________
4. how to use words grammatically	____	____	____________
5. dictionary use	____	____	____________
(specify)			____________________

	yes	no	frequency
20. Do you use dictionaries in the University Library?	____	____	____________
21. Do you use more than one dictionary to look up the same item?	____	____	____________
22. When reading, do you look up all difficult words?	____	____	____________
23. Do you write the looked-up meanings in the text you are reading?	____	____	____________
24. Do you keep a special vocabulary notebook for looked-up items?	____	____	____________
25. Do you refer to the introductory material in dictionaries you use?	____	____	____________
26. Do you browse through dictionary pages without the intention of looking up something in particular?	____	____	____________
27. Do you refer to glossaries in textbooks?	____	____	____________
28. Do you write the pronunciation of English words in Arabic? (examples to be given)	____	____	____________
29. Do you guess the meanings of difficult words before you refer to a dictionary?	____	____	____________

<u>Section Five</u>: Attitudes

30. What do you think of the use of the following types of dictionaries? Rate each type from 1 = not important to 5 = extremely important. (circle number 6 if interviewee does not know)

1. using dictionaries in general	1	2	3	4	5	6
2. medical/nursing large English-Arabic	1	2	3	4	5	6
3. medical/nursing large English-English	1	2	3	4	5	6
4. medical/nursing large Arabic-English	1	2	3	4	5	6
5. medical/nursing pocket English-Arabic	1	2	3	4	5	6
6. medical/nursing pocket English-English	1	2	3	4	5	6
7. medical/nursing pocket Arabic-English	1	2	3	4	5	6
8. general large English-Arabic	1	2	3	4	5	6
9. general large English-English	1	2	3	4	5	6
10. general large Arabic-English	1	2	3	4	5	6
11. general pocket English-Arabic	1	2	3	4	5	6
12. general pocket English-English	1	2	3	4	5	6
13. general pocket Arabic-English	1	2	3	4	5	6

31. What do you think of introducing information and exercises on dictionaries and how to use them? Rate from 1 = unnecessary to 5 = extremely necessary (circle 6 if the interviewee does not know)

31.1 1 2 3 4 5 6

31.2 On what types of dictionaries should the information/ exercises concentrate?

1. medical/nursing English-English	1	2	3	4	5	6
2. medical/nursing English-Arabic	1	2	3	4	5	6
3. medical/nursing Arabic-English	1	2	3	4	5	6
4. general English-English	1	2	3	4	5	6
5. general English-Arabic	1	2	3	4	5	6
6. general Arabic-English	1	2	3	4	5	6

32. What do you think of allowing students to use a dictionary in:

nursing examinations	1 2 3 4 5 6
English language examinations	1 2 3 4 5 6

Section Seven: Needs

33. How much help with English do you need in the following? Rate from 1 = no need to 5 = much need.

1. grammar	1 2 3 4 5
2. pronunciation	1 2 3 4 5
3. meaning	1 2 3 4 5
4. abbreviations	1 2 3 4 5
5. spelling	1 2 3 4 5
6. punctuation	1 2 3 4 5
7. other? (specify) ________________	1 2 3 4 5
8. reading textbooks	1 2 3 4 5
9. reading journals	1 2 3 4 5
10. reading handouts	1 2 3 4 5
11. understanding lectures	1 2 3 4 5
12. understanding seminars	1 2 3 4 5
13. understanding rounds	1 2 3 4 5
14. writing essays	1 2 3 4 5
15. writing reports	1 2 3 4 5
16. writing care plans	1 2 3 4 5
17. speaking in lectures	1 2 3 4 5
18. speaking in seminars	1 2 3 4 5
19. speaking on rounds	1 2 3 4 5
20. giving oral presentations	1 2 3 4 5

APPENDIX 4: THE ESP STAFF INTERVIEW
(implemented in Arabic)

The Language Centre
University of Jordan

ESP Staff Interview

Introduction

1. A brief idea about the purpose of the research and the nature of the interview.

2. Assuring the interviewee about the confidentiality of information provided.

3. Guaranteeing the interviewee's right not to answer any question or part of question s/he may feel sensitive about.

Section One: Background Information

1. Nationality: 1. Jordanian 2. non-Jordanian

   (specify) ________________

2. Years of experience in university teaching: ________________

3. Academic qualifications:

	Degree/Certificate	Awarding Body
1.	__________________	__________________
2.	__________________	__________________
3.	__________________	__________________
4.	__________________	__________________

4. Level/s of students taught:

   1. first year 2. second year

   3. third year 4. fourth year

   4. postgraduates

5. Course/s being taught

   1. 2.

   3. 4.

Section Two: Dictionary Use Before Qualification

6. Before you started teaching at the university level:

   1. Did you use dictionaries? 1. yes 2. no

      (if the answer is no, move to question 7)

   2. How often did you use dictionaries? ________________

   3. What types of dictionaries did you use?

		yes	no
3.1	English-Arabic	___	___
3.2	English-English	___	___
3.3	Arabic-English	___	___

   4. What sizes of dictionaries did you use?

4.1	pocket	___	___
4.2	medium	___	___
4.3	large	___	___

   5. What names of dictionaries did you use?

      5.1 ____________________________

      5.2 ____________________________

      5.3 ____________________________

      5.4 ____________________________

7. Have you received information on dictionaries and training in how to use them?

   1. yes 2. no (if the answer is yes)

   7.1 At what stage/s did you receive such training?

   ____

   7.2 What type of training have your received?

   ____

   7.3 How do you evaluate such training?

   ____

Section Three: In-Service Dictionary Use

8. Do you use dictionaries now? 1. yes 2. no

(if the answer is no, move to question 12/Section Four)

9. What name/s, type/s, size/s of dictionaries do you use?

	dictionary	type	size	frequency
1.	________________	________	________	___________
2.	________________	________	________	___________
3.	________________	________	________	___________
4.	________________	________	________	___________
5.	________________	________	________	___________

10. In what situations do you use dictionaries?

	yes	no	specification
1. reading	___	___	_____________
2. writing	___	___	_____________
3. listening	___	___	_____________
4. speaking	___	___	_____________
5. other	___	___	_____________

11. For what purposes do you use dictionaries

	yes	no	specification
1. meaning	___	___	_____________
2. pronunciation	___	___	_____________
3. spelling	___	___	_____________
4. syllabification	___	___	_____________
5. etymology	___	___	_____________
6. grammar	___	___	_____________
7. other	___	___	_____________

Section Four: Dictionary Use and ESP Teaching

12. What dictionary/ies do you think your students use?

	dictionary	type	size	frequency
1.	________	________	________	________
2.	________	________	________	________
3.	________	________	________	________
4.	________	________	________	________

13. Do you carry a dictionary to your classrooms?

1. yes 2. no (specify ____________________)

14. Do you use a dictionary in preparing your ESP lessons?

1. yes 2. no (specify ____________________)

15. Do you provide your students with information/exercises on dictionaries and how to use them?

1. yes 2. no (specify ____________________)

16. Do you test your students' knowledge of (and ability to use) dictionaries?

1. yes 2. no (specify ____________________)

17. Do you encourage your students to use particular dictionaries?

1. yes 2. no (specify ____________________)

18. In your teaching, do you explain meanings of difficult vocabulary to your students?

1. yes 2. no (specify ____________________)

19. Do your students ask you for help with meanings of words?

1. yes 2. no (specify ____________________)

20. Do you help your students with pronunciation?

1. yes 2. no (specify ____________________)

21. Do your students ask you for help with pronunciation?

1. yes 2. no (specify ____________________)

22. In your opinion, what particular difficulties do your students face with vocabulary?

    1. ______________________________
    2. ______________________________
    3. ______________________________
    4. ______________________________
    5. ______________________________
    6. ______________________________

23. In your opinion, what particular difficulties do your students face with dictionaries?

    1. ______________________________
    2. ______________________________
    3. ______________________________
    4. ______________________________
    5. ______________________________
    6. ______________________________

24. What particular dictionaries would you like to recommend for your students to use?

	dictionary	type	size
1.	__________	______	______
2.	__________	______	______
3.	__________	______	______
4.	__________	______	______

<u>Section Five</u>: Attitudes

25. Please give your opinion concerning the importance you attach to the following (choose from number 1 = not important to 5 = very important).

    1. Students' use of dictionaries

       general 1 2 3 4 5

2. Students' use of the following types of dictionaries:

- general large English-Arabic	1	2	3	4	5
- general large English-English	1	2	3	4	5
- general large Arabic-English	1	2	3	4	5
- general pocket English-Arabic	1	2	3	4	5
- general pocket English-English	1	2	3	4	5
- general pocket Arabic-English	1	2	3	4	5
- technical large English-Arabic	1	2	3	4	5
- technical large English-English	1	2	3	4	5
- technical large Arabic-English	1	2	3	4	5
- technical pocket English-Arabic	1	2	3	4	5
- technical pocket English-English	1	2	3	4	5
- technical pocket Arabic-English	1	2	3	4	5

26. What do you think of allowing students to use dictionaries in:

	agree	disagree	undecided
1. nursing examinations	_____	_____	_____
2. ESP examinations	_____	_____	_____

27. Would you like to offer any general comments/suggestions concerning dictionary use and/or vocabulary teaching in ESP for nursing students?

Thank you very much for your help and for your time.

APPENDIX 5: TABLES FOR FURTHER REFERENCE

Table 1 : First-Year Students' Strategies (N = 127)

Strategy		Frequency			Total
		low	average	high	
Guessing from Context in Reading	n %	4 3.3	31 25.8	85 70.8	120 94.4
Glossing in Text	n %	4 3.3	17 14.2	98 82.3	119 93.7
Asking ESP Staff about Meaning	n %	33 29.7	30 27.0	48 43.2	111 87.4
Asking fellow Students about Meaning	n %	15 13.7	41 37.6	53 48.6	109 85.8
Asking ESP Staff about Pronunciation	n %	34 31.2	31 28.4	44 40.3	109 85.8
Asking fellow Students about Pronunciation	n %	35 32.1	36 33.0	38 34.8	109 85.8
Asking Nursing Staff about Meaning	n %	29 27.1	44 41.1	34 31.7	107 84.2
Browsing through Dictionary Pages	n %	12 11.6	45 43.6	46 44.6	103 81.1
Looking up all Items in Reading	n %	14 13.8	27 26.7	60 59.4	101 79.5
Asking Nursing Staff about Pronunciation	n %	34 34.7	40 40.8	24 24.4	98 77.1
Consulting Textbook Glossaries	n %	23 24.7	35 37.6	35 37.6	93 73.2
Glossing in a special Vocabulary Notebook	n %	20 28.5	24 34.2	26 37.1	70 55.1
Looking up the same Item in more than one Dictionary	n %	24 34.7	24 34.7	21 30.4	69 54.3
Asking ESP Staff about Type of Dictionary	n %	28 41.1	22 32.3	18 26.4	68 53.5
Asking ESP Staff about Symbols in Dictionaries	n %	23 33.8	25 36.7	20 29.4	68 53.5
Using Dictionaries in University Library	n %	20 30.3	29 43.9	17 25.7	66 51.9
Consulting Dictionary Front Matter	n %	28 44.4	18 28.5	17 26.9	63 49.6
Asking fellow Students about Type of Dictionary	n %	24 40.0	16 26.6	20 33.3	60 47.2
Asking Nursing Staff about Symbols in Dictionaries	n %	20 34.4	21 36.2	17 29.3	58 45.6
Transliteration	n %	21 36.2	21 36.2	16 27.5	58 43.7
Asking ESP Staff about how to use Dictionaries	n %	23 41.8	22 40.0	10 18.1	55 43.3
Asking Nursing Staff about Type of Dictionary	n %	14 28.0	12 24.0	24 48.0	50 39.3
Asking fellow Students about Symbols in Dictionaries	n %	19 38.7	20 40.8	10 20.4	49 38.5
Asking fellow Students about how to use Dictionaries	n %	16 50.0	6 18.7	10 31.2	32 25.2
Asking Nursing Staff about how to use Dictionaries	n %	14 46.6	10 33.3	6 20.0	30 23.6

Table 2 : Second-Year Students' Strategies (N = 96)

Strategy		Frequency			Total
		low	average	high	
Guessing from Context in Reading	n	13	24	53	90
	%	14.4	26.6	58.8	93.7
Glossing in Text	n	10	18	62	90
	%	11.1	20.0	68.8	93.7
Asking fellow Students about Meaning	n	17	33	39	89
	%	19.1	37.0	43.8	92.7
Asking Nursing Staff about Meaning	n	14	36	35	85
	%	16.4	42.3	41.1	88.5
Asking fellow Students about Pronunciation	n	19	30	31	80
	%	23.7	37.5	38.7	83.3
Looking up all Items in Reading	n	18	26	35	79
	%	22.7	32.9	44.3	82.3
Browsing through Dictionary Pages	n	23	31	24	78
	%	29.4	39.7	30.7	81.2
Asking ESP Staff about Meaning	n	24	28	25	77
	%	31.1	36.3	32.4	80.2
Asking ESP Staff about Pronunciation	n	28	26	20	74
	%	37.8	35.1	27.0	77.0
Consulting Textbook Glossaries	n	23	25	25	73
	%	31.5	34.2	34.2	76.0
Asking Nursing Staff about Pronunciation	n	29	26	17	72
	%	40.2	36.1	23.6	75.0
Looking up the same item in more than one Dictionary	n	17	29	17	63
	%	26.9	46.0	26.9	65.6
Asking Nursing Staff about Type of Dictionary	n	26	19	12	57
	%	45.6	33.3	21.0	59.3
Asking fellow students about Symbols in Dictionaries	n	28	15	14	57
	%	49.1	26.3	24.5	59.3
Asking fellow students about Type of Dictionary	n	23	13	18	54
	%	42.6	24.0	33.3	56.2
Consulting Dictionary Front Matter	n	24	9	18	51
	%	47.0	17.6	35.3	53.1
Asking ESP Staff about Type of Dictionary	n	23	12	13	48
	%	47.9	25.0	27.0	50.0
Using Dictionaries in University Library	n	19	18	10	47
	%	40.4	38.3	21.2	48.9
Asking fellow Students about how to use Dictionaries	n	18	11	16	45
	%	40.0	24.4	35.5	46.8
Asking ESP Staff about Symbols in Dictionaries	n	22	10	10	42
	%	52.3	23.8	23.8	43.7
Asking Nursing Staff about Symbols in Dictionaries	n	16	18	8	42
	%	38.0	42.8	19.0	43.7
Transliteration	n	17	19	6	42
	%	40.4	45.2	14.2	43.7
Glossing in a special Vocabulary Notebook	n	16	11	12	39
	%	41.0	28.2	30.7	40.6
Asking ESP Staff about how to use Dictionaries	n	17	9	9	35
	%	48.5	25.7	25.7	36.4

Table 3 : Third-Year Students' Strategies (N = 79)

Strategy		Frequency			Total
		low	average	high	
Guessing from Context in Reading	n	3	22	53	78
	%	3.8	28.2	67.9	98.7
Asking fellow Students about Meaning	n	12	22	44	78
	%	15.3	28.2	56.4	98.7
Asking Nursing Staff about Meaning	n	12	30	35	77
	%	15.5	38.9	45.4	97.4
Asking fellow Students about Pronunciation	n	22	24	30	76
	%	38.9	31.5	39.4	96.2
Glossing in Text	n	4	25	45	74
	%	5.4	33.7	60.8	93.6
Asking ESP Staff about Meaning	n	11	36	26	73
	%	15.0	49.3	35.6	92.4
Looking up all Items in Reading	n	17	33	23	73
	%	23.2	45.2	31.5	92.4
Asking ESP Staff about Pronunciation	n	11	34	24	69
	%	15.9	49.2	34.7	87.3
Asking Nursing Staff about Pronunciation	n	21	32	15	68
	%	30.8	47.0	22.0	86.0
Consulting Textbook Glossaries	n	17	19	27	63
	%	26.9	30.1	42.8	79.7
Browsing through Dictionary Pages	n	17	22	20	59
	%	28.8	37.2	33.9	74.6
Asking ESP Staff about Symbols in Dictionaries	n	23	15	8	46
	%	50.0	32.5	17.4	58.2
Asking ESP Staff about Type of Dictionary	n	23	15	7	45
	%	51.1	33.3	15.5	56.9
Asking Nursing Staff about Symbols in Dictionaries	n	23	15	5	43
	%	53.4	34.8	11.6	54.4
Asking fellow Students about Type of Dictionary	n	18	18	4	40
	%	45.0	45.0	10.0	50.6
Asking fellow Students about Symbols in Dictionaries	n	20	11	9	40
	%	50.0	27.5	22.5	50.6
Consulting Dictionary Front Matter	n	15	15	10	40
	%	37.5	37.5	25.0	50.6
Using Dictionaries in University Library	n	24	11	4	39
	%	61.5	28.2	10.2	49.3
Glossing in a special Vocabulary Notebook	n	15	12	10	37
	%	40.5	32.4	27.0	46.8
Asking ESP Staff about how to use Dictionaries	n	16	15	5	36
	%	44.4	41.6	13.8	45.5
Looking up the same Item in more than one Dictionary	n	15	12	6	33
	%	45.4	36.3	18.1	41.7
Asking Nursing Staff about Type of Dictionary	n	16	13	2	31
	%	51.6	41.9	6.4	39.2
Asking fellow Students about how to use Dictionaries	n	12	8	5	25
	%	48.0	32.0	20.0	31.6
Asking Nursing Staff about how to use Dictionaries	n	12	7	5	24
	%	50.0	29.[illegible]	20.8	30.3
Transliteration	n	10	12	1	23
	%	43.4	52.[illegible]	4.3	29.1

Table 4 : Fourth-Year Students' Strategies (N = 102)

Strategy		Frequency			Total
		low	average	high	
Asking fellow Students about Meaning	n	18	40	42	100
	%	18.0	40.0	42.0	98.0
Asking Nursing Staff about Meaning	n	20	41	37	98
	%	20.4	41.8	37.7	96.0
Guessing from Context in Reading	n	3	28	67	98
	%	3.0	28.5	68.3	96.0
Asking ESP Staff about Meaning	n	19	37	36	92
	%	20.6	40.2	39.1	90.2
Glossing in Text	n	15	22	55	92
	%	16.3	23.9	59.7	90.2
Looking up all Items in Reading	n	23	32	36	91
	%	25.2	35.1	39.5	89.2
Asking fellow Students about Pronunciation	n	32	32	24	88
	%	36.3	36.3	27.2	86.2
Asking ESP Staff about Pronunciation	n	28	27	31	86
	%	32.5	31.4	36.0	84.3
Consulting Textbook Glossaries	n	20	27	38	85
	%	23.5	31.7	44.7	83.3
Asking Nursing Staff about Pronunciation	n	33	30	18	81
	%	40.7	37.0	27.2	79.4
Browsing through Dictionary Pages	n	17	38	23	78
	%	21.8	48.7	29.4	76.4
Using Dictionaries in University Library	n	24	24	13	61
	%	39.3	39.3	21.3	59.8
Consulting Dictionary Front Matter	n	12	19	22	53
	%	22.6	35.8	41.5	51.9
Asking fellow Students about Type of Dictionary	n	21	20	11	52
	%	40.3	38.4	21.1	50.9
Asking ESP Staff about Type of Dictionary	n	23	16	12	51
	%	45.1	31.3	23.5	50.5
Looking up the same Item in more than one Dictionary	n	20	19	10	49
	%	40.8	38.7	20.4	48.0
Asking fellow Students about Symbols in Dictionaries	n	23	19	6	48
	%	47.9	39.5	12.5	47.0
Asking ESP Staff about Symbols in Dictionaries	n	20	13	11	44
	%	45.4	29.5	25.0	43.1
Asking ESP Staff about how to use Dictionaries	n	17	12	9	38
	%	44.7	31.5	23.6	37.2
Asking fellow Students about how to use Dictionaries	n	22	11	4	37
	%	59.4	29.7	10.8	36.2
Glossing in a special Vocabulary Notebook	n	19	8	9	36
	%	52.7	22.2	25.0	35.3
Asking Nursing Staff about Symbols in Dictionaries	n	17	13	4	34
	%	50.0	38.2	11.7	33.3
Asking Nursing Staff about Type of Dictionary	n	20	6	7	33
	%	60.6	18.1	21.2	32.3
Transliteration	n	18	8	1	27
	%	66.6	29.6	3.7	26.4
Asking Nursing Staff about how to use Dictionaries	n	15	7	1	23
	%	65.2	30.4	4.3	22.5

Table 5 : Students' Strategies : Total Sample (N = 405)

Strategy		Frequency low	average	high	Total
Guessing from Context in Reading	n	23	106	258	387
	%	5.9	27.4	66.6	95.5
Asking fellow Students about Meaning	n	62	136	179	377
	%	16.4	36.0	47.4	93.0
Glossing in Text	n	33	82	261	376
	%	8.7	21.8	69.4	92.8
Asking Nursing Staff about Meaning	n	75	152	141	368
	%	20.3	41.3	38.3	90.8
Asking ESP Staff about Meaning	n	88	131	135	354
	%	24.8	37.0	38.1	87.4
Asking fellow Students about Pronunciation	n	108	123	123	354
	%	30.5	34.7	34.7	87.4
Looking up all Items in Reading	n	72	118	155	345
	%	20.8	34.2	44.9	85.1
Asking ESP Staff about Pronunciation	n	102	118	119	339
	%	30.0	34.8	35.1	83.7
Asking Nursing Staff about Pronunciation	n	117	129	74	320
	%	36.5	40.3	23.1	79.0
Browsing through Dictionary Pages	n	69	136	113	318
	%	21.7	42.7	35.5	78.5
Consulting Textbook Glossaries	n	83	107	125	315
	%	26.3	33.9	39.6	77.7
Looking up the same Item in more than one Dictionary	n	77	84	54	215
	%	35.8	39.0	25.1	53.0
Asking ESP Staff about Type of Dictionary	n	98	65	50	213
	%	46.0	30.5	23.4	52.6
Using Dictionaries in University Library	n	87	82	44	213
	%	40.8	38.5	20.6	52.6
Consulting Dictionary Front Matter	n	80	61	67	208
	%	38.4	29.3	32.2	51.3
Asking fellow Students about Type of Dictionary	n	86	68	53	207
	%	41.5	32.8	25.6	51.1
Asking ESP Staff about Symbols in Dictionaries	n	89	63	49	201
	%	44.2	31.3	24.3	49.6
Asking fellow Students about Symbols in Dictionaries	n	90	65	39	194
	%	46.4	33.5	20.1	47.9
Glossing in a special Vocabulary Notebook	n	71	55	57	183
	%	38.8	30.0	31.1	45.1
Asking Nursing Staff about Symbols in Dictionaries	n	76	67	34	177
	%	42.9	37.8	19.2	43.7
Asking ESP Staff about how to use Dictionaries	n	74	58	33	165
	%	44.8	35.1	20.0	40.7
Asking Nursing Staff about Type of Dictionary	n	77	50	45	172
	%	44.7	29.0	26.1	42.4
Transliteration	n	66	60	24	150
	%	44.0	40.0	16.0	37.0
Asking fellow Students about how to use Dictionaries	n	68	36	35	139
	%	48.9	25.9	25.1	34.3
Asking Nursing Staff about how to use Dictionaries	n	54	35	19	108
	%	50.0	32.4	17.6	26.6

Table 6 : First-Year Students (N = 127) : Needs Analysis

Context of Language Use		Need for help			Total
		little	average	much	
Understanding Lectures	n	27	27	55	109
	%	24.7	24.7	50.4	85.8
Understanding Oral Exam Questions	n	22	30	61	113
	%	19.4	26.5	53.9	88.9
Understanding Assigned Textbooks	n	24	37	53	114
	%	21.0	32.4	46.4	89.7
Understanding Handouts	n	28	46	35	109
	%	25.6	42.2	32.1	85.8
Understanding Specialist Journals	n	33	39	40	112
	%	29.4	34.8	35.7	88.1
Understanding Written Exam Questions	n	28	34	50	112
	%	25.0	30.3	44.6	88.1
Writing Written Exam Answers	n	26	29	52	107
	%	24.2	27.1	48.5	84.2
Writing Notes when Listening	n	28	36	44	108
	%	25.9	33.3	40.7	85.0
Speaking in Lectures	n	21	23	70	114
	%	18.4	20.1	61.4	89.7
Answering Oral Exam Questions	n	21	22	72	115
	%	18.2	19.1	62.6	90.5
Practising Speaking on Nursing Topics	n	8	20	95	123
	%	6.5	16.2	77.2	96.8
Practising Writing on Nursing Topics	n	7	32	84	123
	%	5.6	26.0	68.2	96.8
Practising Listening Comprehension	n	12	33	76	121
on Nursing Topics	%	9.9	27.2	62.8	95.2
Practising Reading Comprehension	n	13	35	71	119
on Nursing Topics	%	10.9	29.4	59.6	93.7
Practice in Grammar	n	26	40	46	112
	%	23.2	35.7	41.0	88.1
Practice in Pronunciation	n	24	40	58	122
	%	19.6	32.7	47.5	96.0
Practice in Vocabulary Meaning	n	18	35	63	116
	%	15.5	30.1	54.3	91.3
Practice in Common Nursing	n	15	33	75	123
Abbreviations	%	12.1	26.8	60.9	96.8
Practice in Spelling	n	21	32	70	123
	%	17.0	26.0	56.9	96.8
Practice in Punctuation	n	28	36	41	105
	%	26.6	34.2	39.0	82.6
Practice in Using Monolingual	n	29	36	29	94
English Dictionaries	%	30.8	38.2	30.8	74.0

Table 7 : Second-Year Students (N = 96) : Needs Analysis

Context of Language Use		Need for help			Total
		little	average	much	
Understanding Lectures	n	24	21	36	81
	%	29.6	25.9	44.4	84.3
Understanding Seminars	n	21	25	36	82
	%	25.6	30.4	43.9	85.4
Understanding Clinical Rounds	n	26	18	43	87
	%	29.8	20.6	49.4	90.6
Understanding Oral Exam Questions	n	16	17	49	82
	%	19.5	20.7	59.7	85.4
Understanding Assigned Textbooks	n	21	30	38	89
	%	23.5	33.7	42.6	92.7
Understanding Handouts	n	32	40	16	88
	%	36.3	45.4	18.1	91.6
Understanding Specialist Journals	n	28	30	29	87
	%	32.1	34.4	33.3	90.6
Understanding Written Exam Questions	n	21	24	42	87
	%	24.1	27.5	48.2	90.6
Writing Research Essays	n	16	24	48	88
	%	18.1	27.2	54.5	91.6
Writing Technical Reports	n	15	27	43	85
	%	17.6	31.7	50.5	88.5
Writing Nursing Care Plans	n	17	20	40	77
	%	22.0	25.9	51.9	80.2
Writing Written Exam Answers	n	20	27	35	82
	%	24.3	32.9	42.6	85.4
Writing Notes when Listening	n	17	31	32	80
	%	21.2	38.7	40.0	83.3
Speaking in Lectures	n	17	21	51	89
	%	19.1	23.5	57.3	92.7
Speaking in Seminars	n	16	27	44	87
	%	18.3	31.0	50.5	90.6
Speaking in Clinical Rounds	n	18	22	48	88
	%	20.4	25.0	54.5	91.6
Giving Oral Presentations	n	12	25	49	86
	%	13.9	29.0	56.9	89.5
Answering Oral Exam Questions	n	10	31	45	86
	%	11.6	36.0	52.3	89.5
Practising Speaking on Nursing Topics	n	15	20	56	91
	%	16.4	21.9	61.5	94.7
Practising Writing on Nursing Topics	n	13	32	45	90
	%	14.4	35.5	50.0	93.7
Practising Listening Comprehension on Nursing Topics	n	13	31	46	90
	%	14.4	34.4	51.1	93.7
Practising Reading Comprehension on Nursing Topics	n	17	27	43	87
	%	19.5	31.0	49.4	90.6
Practice in Grammar	n	26	26	38	90
	%	28.8	28.8	42.2	93.7
Practice in Pronunciation	n	24	29	34	87
	%	27.5	33.3	39.0	90.6
Practice in Vocabulary Meaning	n	26	25	42	93
	%	27.9	26.8	45.1	96.8
Practice in Common Nursing Abbreviations	n	17	20	48	85
	%	20.0	23.5	56.4	88.5
Practice in Spelling	n	21	30	36	87
	%	24.1	34.4	41.3	90.6
Practice in Punctuation	n	23	26	28	77
	%	29.8	33.7	36.3	80.2
Practice in Using Monolingual English Dictionaries	n	16	24	32	72
	%	22.2	33.3	44.4	75.0

Table 8 : Third-Year Students (N = 79) : Needs Analysis

Context of Language Use		Need for help: little	average	much	Total
Understanding Lectures	n	25	15	29	69
	%	36.2	21.7	42.0	87.3
Understanding Seminars	n	20	24	28	72
	%	27.7	33.3	38.8	91.1
Understanding Clinical Rounds	n	19	21	32	72
	%	26.3	29.1	44.4	91.1
Understanding Oral Exam Questions	n	13	19	32	64
	%	20.3	29.6	50.0	81.0
Understanding Assigned Textbooks	n	20	23	32	75
	%	26.6	30.6	42.6	94.9
Understanding Handouts	n	27	31	21	79
	%	34.1	39.2	26.5	100
Understanding Specialist Journals	n	16	34	24	74
	%	21.6	45.9	32.4	93.6
Understanding Written Exam Questions	n	22	14	35	71
	%	30.9	19.7	49.2	89.8
Writing Research Essays	n	18	17	41	76
	%	23.6	22.3	53.9	96.2
Writing Technical Reports	n	16	20	36	72
	%	22.2	27.7	50.0	91.1
Writing Nursing Care Plans	n	15	14	33	62
	%	24.1	22.5	53.2	78.4
Writing Written Exam Answers	n	21	16	29	66
	%	31.8	24.2	43.9	83.5
Writing Notes when Listening	n	19	18	29	66
	%	28.7	27.2	43.9	83.5
Speaking in Lectures	n	18	18	38	74
	%	24.3	24.3	51.3	93.6
Speaking in Seminars	n	18	22	34	74
	%	24.3	29.2	45.9	93.6
Speaking in Clinical Rounds	n	14	26	33	73
	%	19.1	35.6	45.2	92.4
Giving Oral Presentations	n	16	19	39	74
	%	21.6	25.6	52.7	93.6
Answering Oral Exam Questions	n	17	20	33	70
	%	24.2	28.5	47.1	88.6
Practising Speaking on Nursing Topics	n	11	9	54	74
	%	14.8	12.1	72.9	93.6
Practising Writing on Nursing Topics	n	12	23	40	75
	%	16.0	30.6	53.3	94.9
Practising Listening Comprehension on Nursing Topics	n	16	21	39	76
	%	21.0	27.6	51.3	96.2
Practising Reading Comprehension on Nursing Topics	n	17	19	35	71
	%	23.9	26.7	49.2	89.8
Practice in Grammar	n	15	19	37	71
	%	21.1	26.7	52.1	89.8
Practice in Pronunciation	n	22	22	34	78
	%	28.2	28.2	43.5	98.7
Practice in Vocabulary Meaning	n	23	22	32	77
	%	29.8	28.7	41.5	97.4
Practice in Common Nursing Abbreviations	n	16	19	35	70
	%	22.8	27.1	50.0	88.6
Practice in Spelling	n	19	17	37	73
	%	26.0	23.2	50.6	92.4
Practice in Punctuation	n	14	30	24	68
	%	20.5	44.1	35.2	86.0
Practice in Using Monolingual English Dictionaries	n	18	23	18	59
	%	30.5	38.9	30.5	74.6

Table 9 : Fourth-Year Students (N = 102) : Needs Analysis

Context of Language Use		Need for help			Total
		little	average	much	
Understanding Lectures	n	27	24	24	75
	%	36.0	32.0	32.0	73.5
Understanding Seminars	n	30	30	24	84
	%	35.7	35.7	28.5	82.3
Understanding Clinical Rounds	n	32	30	21	83
	%	38.5	36.1	25.3	81.3
Understanding Oral Exam Questions	n	21	22	38	81
	%	25.9	27.1	46.9	79.4
Understanding Assigned Textbooks	n	37	28	26	91
	%	40.6	30.7	28.5	89.2
Understanding Handouts	n	30	42	19	91
	%	32.9	46.1	20.8	89.2
Understanding Specialist Journals	n	29	30	31	90
	%	32.2	33.3	34.0	88.2
Understanding Written Exam Questions	n	31	28	25	84
	%	36.9	33.3	29.7	82.3
Writing Research Essays	n	20	20	50	90
	%	22.2	22.2	55.5	88.2
Writing Technical Reports	n	23	27	32	82
	%	28.0	32.9	39.0	80.3
Writing Nursing Care Plans	n	24	21	25	70
	%	34.2	30.0	35.7	68.6
Writing Written Exam Answers	n	32	23	28	83
	%	38.5	27.7	33.7	81.3
Writing Notes when Listening	n	28	26	24	78
	%	35.8	33.3	30.7	76.4
Speaking in Lectures	n	20	19	48	87
	%	22.9	21.8	55.1	85.2
Speaking in Seminars	n	15	22	48	85
	%	17.6	25.8	56.4	83.3
Speaking in Clinical Rounds	n	24	27	34	85
	%	28.2	31.7	40.0	83.3
Giving Oral Presentations	n	21	25	40	86
	%	24.4	29.0	46.5	84.3
Answering Oral Exam Questions	n	21	25	39	85
	%	24.7	29.4	45.8	83.3
Practising Speaking on Nursing Topics	n	14	31	46	91
	%	15.3	34.0	50.5	89.2
Practising Writing on Nursing Topics	n	21	32	33	86
	%	24.4	37.2	38.3	84.3
Practising Listening Comprehension on Nursing Topics	n	28	24	36	88
	%	31.8	27.2	40.6	86.2
Practising Reading Comprehension on Nursing Topics	n	30	28	28	86
	%	34.8	32.5	32.5	84.3
Practice in Grammar	n	19	27	44	90
	%	21.1	30.0	48.8	88.2
Practice in Pronunciation	n	27	32	30	89
	%	30.3	35.9	33.7	87.2
Practice in Vocabulary Meaning	n	30	28	31	89
	%	33.7	31.4	34.8	87.2
Practice in Common Nursing Abbreviations	n	26	15	41	82
	%	31.7	18.2	50.0	80.3
Practice in Spelling	n	22	24	38	84
	%	26.1	28.5	45.2	82.3
Practice in Punctuation	n	22	25	38	85
	%	25.8	29.4	44.7	83.3
Practice in Using Monolingual English Dictionaries	n	24	27	25	76
	%	31.5	35.5	32.8	74.5

Table 10 : Total Sample Students (N = 405) : Needs Analysis

Context of Language Use		Need for help			Total
		little	average	much	
Understanding Lectures	n	103	88	144	335
	%	30.7	26.2	42.9	82.7
Understanding Seminars (first-year excluded)	n	71	79	88	238
	%	29.8	33.1	36.9	85.6
Understanding Clinical Rounds (first-year excluded)	n	77	69	96	242
	%	31.8	28.5	39.6	87.0
Understanding Oral Exam Questions	n	72	88	181	341
	%	21.1	25.8	53.0	84.1
Understanding Assigned Textbooks	n	102	119	149	370
	%	27.5	32.1	40.2	91.3
Understanding Handouts	n	117	160	91	368
	%	31.7	43.4	24.7	90.8
Understanding Specialist Journals	n	106	134	124	364
	%	29.1	36.8	34.0	89.8
Understanding Written Exam Questions	n	102	101	152	355
	%	28.7	28.4	42.8	87.6
Writing Research Essays (first-year excluded)	n	54	61	139	254
	%	21.2	24.1	54.7	91.3
Writing Technical Reports (first-year excluded)	n	54	74	111	239
	%	22.6	30.9	46.4	85.9
Writing Nursing Care Plans (first-year excluded)	n	56	55	98	209
	%	26.8	26.3	46.8	75.1
Writing Written Exam Answers	n	99	96	144	339
	%	29.2	28.3	42.4	83.7
Writing Notes when Listening	n	93	111	129	333
	%	27.9	33.3	38.7	82.2
Speaking in Lectures	n	76	81	208	365
	%	20.8	22.1	56.9	90.1
Speaking in Seminars (first-year excluded)	n	49	71	126	246
	%	19.9	28.8	51.2	88.4
Speaking in Clinical Rounds (first-year excluded)	n	56	75	115	246
	%	22.7	30.4	46.7	88.4
Giving Oral Presentations (first-year excluded)	n	49	69	128	246
	%	19.9	28.0	52.0	88.4
Answering Oral Exam Questions	n	69	99	189	357
	%	19.3	27.7	52.9	88.1
Practising Speaking on Nursing Topics	n	48	80	252	380
	%	12.6	21.0	66.3	93.8
Practising Writing on Nursing Topics	n	53	119	203	375
	%	14.1	31.7	54.1	92.5
Practising Listening Comprehension on Nursing Topics	n	69	109	198	376
	%	18.3	28.9	52.6	92.8
Practising Reading Comprehension on Nursing Topics	n	77	110	177	364
	%	21.1	30.2	48.6	89.8
Practice in Grammar	n	86	113	165	364
	%	23.6	31.0	45.3	89.8
Practice in Pronunciation	n	97	123	157	377
	%	25.7	32.6	41.6	93.0
Practice in Vocabulary Meaning	n	97	110	169	376
	%	25.7	29.2	44.9	92.8
Practice in Common Nursing Abbreviations	n	74	87	200	361
	%	20.4	24.0	55.4	89.1
Practice in Spelling	n	83	104	181	368
	%	22.5	28.2	49.1	90.8
Practice in Punctuation	n	87	117	132	336
	%	25.8	34.8	39.2	82.9
Practice in Using Monolingual English Dictionaries	n	88	110	104	302
	%	29.1	36.4	34.4	74.5

Table 11 : First-Year Students' Attitudes (N = 127)

Statement		Attitude		
		agree	disagree	undecided
"It is important for nursing students to use dictionaries in their university study"	n %	110 86.6	13 10.2	4 3.1
"In general, dictionary using is boring"	n %	67 52.7	54 42.5	6 4.7
"Nursing students need to refer to a dictionary when they write in English"	n %	78 61.4	37 29.1	12 9.4
"To depend on one dictionary is not enough"	n %	75 59.0	35 27.5	17 13.4
"Nursing students can understand the content of their assigned textbooks without reference to a dictionary"	n %	26 20.4	88 69.3	13 10.2
"Nursing students should be allowed to use dictionaries during nursing exams"	n %	64 50.4	56 44.1	7 5.5
"Nursing students should be allowed to use dictionaries during English language exams"	n %	54 42.5	64 50.4	9 7.1
"Every nursing student should have at least one dictionary"	n %	117 92.1	7 5.5	3 2.3
"It is better for nursing students to learn English through nursing-related texts/subjects"	n %	94 74.0	24 18.9	9 7.1
"The present English language courses provide sufficient information on dictionaries and how to use them"	n %	24 18.9	68 53.5	35 27.5
"Nursing students should avoid using pocket dictionaries in their university study"	n %	26 20.4	73 57.4	28 22.0
"It is important for nursing students to use a medical/nursing dictionary in their university study"	n %	105 82.6	14 11.0	8 6.3
"English-English dictionaries are more useful to nursing students than English-Arabic dictionaries"	n %	52 40.9	57 44.8	18 14.1
"AL-MAWRID is the most useful English-Arabic dictionary for nursing students"	n %	70 55.1	16 12.6	41 32.2
"In general, studying nursing is enjoyable:	n %	102 80.3	15 11.8	10 7.8
"It would be better if the present English language courses include exercises on English-English dictionaries and how to use them"	n %	98 77.1	18 14.1	11 8.6

Table 12 : Second-Year Students' Attitudes (N = 96)

Statement		Attitude		
		agree	disagree	undecided
"It is important for nursing students to use dictionaries in their university study"	n	83	8	5
	%	86.4	8.3	5.2
"In general, dictionary using is boring"	n	51	35	10
	%	53.1	36.4	10.4
"Nursing students need to refer to a dictionary when they write in English"	n	65	18	13
	%	67.7	18.7	13.5
"To depend on one dictionary is not enough"	n	61	28	7
	%	63.5	29.1	7.3
"Nursing students can understand the content of their assigned textbooks without reference to a dictionary"	n	28	58	10
	%	29.1	60.4	10.4
"Nursing students should be allowed to use dictionaries during <u>nursing</u> exams"	n	52	39	5
	%	54.1	40.6	5.2
"Nursing students should be allowed to use dictionaries during <u>English language</u> exams"	n	41	48	7
	%	42.7	50.0	7.3
"Every nursing student should have at least one dictionary"	n	87	6	3
	%	90.6	6.2	3.1
"It is better for nursing students to learn English through nursing-related texts/subjects"	n	56	31	9
	%	58.3	32.3	9.3
"The present English language courses provide sufficient information on dictionaries and how to use them"	n	11	74	11
	%	11.4	77.0	11.4
"Nursing students should avoid using pocket dictionaries in their university study"	n	15	61	20
	%	15.6	63.5	20.8
"It is important for nursing students to use a medical/nursing dictionary in their university study"	n	82	8	6
	%	85.4	8.3	6.2
"English-English dictionaries are more useful to nursing students than English-Arabic dictionaries"	n	46	29	21
	%	47.9	30.2	21.8
"AL-MAWRID is the most useful English-Arabic dictionary for nursing students"	n	59	12	25
	%	61.4	12.5	26.0
"In general, studying nursing is enjoyable"	n	77	13	6
	%	80.2	13.5	6.2
"It would be better if the present English language courses include exercises on English-English dictionaries and how to use them"	n	76	14	6
	%	79.1	14.5	6.2

Table 13 : Third-Year Students' Attitudes (N = 79)

Statement		Attitude		
		agree	disagree	undecided
"It is important for nursing students to use dictionaries in their university study"	n %	74 93.6	2 2.5	3 3.8
"In general, dictionary using is boring"	n %	31 39.2	37 46.8	11 13.9
"Nursing students need to refer to a dictionary when they write in English"	n %	47 59.5	25 31.6	7 8.8
"To depend on one dictionary is not enough"	n %	33 41.7	36 45.5	10 12.6
"Nursing students can understand the content of their assigned textbooks without reference to a dictionary"	n %	30 37.9	44 55.7	5 6.3
"Nursing students should be allowed to use dictionaries during <u>nursing</u> exams"	n %	31 39.2	40 50.6	8 10.1
"Nursing students should be allowed to use dictionaries during <u>English language</u> exams"	n %	34 43.0	39 49.3	6 7.6
"Every nursing student should have at least one dictionary"	n %	71 89.8	6 7.6	2 2.5
"It is better for nursing students to learn English through nursing-related texts/subjects"	n %	55 69.6	19 24.0	5 6.3
"The present English language courses provide sufficient information on dictionaries and how to use them"	n %	11 13.9	61 77.2	7 8.8
"Nursing students should avoid using pocket dictionaries in their university study"	n %	9 11.4	56 70.8	14 17.7
"It is important for nursing students to use a medical/nursing dictionary in their university study"	n %	71 89.8	7 8.8	1 1.2
"English-English dictionaries are more useful to nursing students than English-Arabic dictionaries"	n %	47 59.5	24 30.3	8 10.1
"AL-MAWRID is the most useful English-Arabic dictionary for nursing students"	n %	61 77.2	7 8.8	11 13.9
"In general, studying nursing is enjoyable"	n %	68 86.0	10 12.6	1 1.2
"It would be better if the present English language courses include exercises on English-English dictionaries and how to use them"	n %	64 81.0	12 15.1	3 3.8

Table 14 : Fourth-Year Students' Attitudes (N = 102)

Statement		Attitude		
		agree	disagree	undecided
"It is important for nursing students to use dictionaries in their university study"	n %	94 92.1	7 6.8	1 0.9
"In general, dictionary using is boring"	n %	35 34.3	61 59.8	6 5.8
"Nursing students need to refer to a dictionary when they write in English"	n %	68 66.6	25 24.5	9 8.8
"To depend on one dictionary is not enough"	n %	53 51.9	35 34.3	14 13.7
"Nursing students can understand the content of their assigned textbooks without reference to a dictionary"	n %	25 24.5	73 71.5	4 3.9
"Nursing students should be allowed to use dictionaries during nursing exams"	n %	42 41.1	56 54.9	4 3.9
"Nursing students should be allowed to use dictionaries during English language exams"	n %	34 33.3	59 57.8	9 8.8
"Every nursing student should have at least one dictionary"	n %	89 87.2	7 6.8	6 5.8
"It is better for nursing students to learn English through nursing-related texts/subjects"	n %	68 66.6	28 27.4	6 5.8
"The present English language courses provide sufficient information on dictionaries and how to use them"	n %	10 9.8	76 74.5	16 15.7
"Nursing students should avoid using pocket dictionaries in their university study"	n %	13 12.7	76 74.5	13 12.7
"It is important for nursing students to use a medical/nursing dictionary in their university study"	n %	91 89.2	7 6.8	4 3.9
"English-English dictionaries are more useful to nursing students than English-Arabic dictionaries"	n %	63 61.7	33 32.3	6 5.8
"AL-MAWRID is the most useful English-Arabic dictionary for nursing students"	n %	70 68.6	14 13.7	18 17.6
"In general, studying nursing is enjoyable"	n %	85 83.3	13 12.7	4 3.9
"It would be better if the present English language courses include exercises on English-English dictionaries and how to use them"	n %	84 82.3	9 8.8	9 8.8

Table 15 : Students' Attitudes (Total Sample) (N = 405)

Statement		Attitude		
		agree	disagree	undecided
"It is important for nursing students to use dictionaries in their university study"	n %	362 89.3	30 7.4	13 3.2
"In general, dictionary using is boring"	n %	184 45.4	187 46.1	34 8.4
"Nursing students need to refer to a dictionary when they write in English"	n %	259 63.9	105 25.9	41 10.1
"To depend on one dictionary is not enough"	n %	223 55.0	134 33.0	48 11.8
"Nursing students can understand the content of their assigned textbooks without reference to a dictionary"	n %	109 26.9	264 65.1	32 7.9
"Nursing students should be allowed to use dictionaries during nursing exams"	n %	190 46.9	191 47.1	24 5.9
"Nursing students should be allowed to use dictionaries during English language exams"	n %	164 40.5	210 51.8	31 7.6
"Every nursing student should have at least one dictionary"	n %	365 90.1	26 6.4	14 3.4
"It is better for nursing students to learn English through nursing-related texts/subjects"	n %	274 67.6	102 25.1	29 7.1
"The present English language courses provide sufficient information on dictionaries and how to use them"	n %	56 13.8	280 69.1	69 17.0
"Nursing students should avoid using pocket dictionaries in their university study"	n %	64 15.8	266 65.6	75 18.5
"It is important for nursing students to use a medical/nursing dictionary in their university study"	n %	350 86.4	36 8.8	19 4.7
"English-English dictionaries are more useful to nursing students than English-Arabic ones"	n %	209 51.6	143 35.3	53 13.0
"AL-MAWRID is the most useful English-Arabic dictionary for nursing students"	n %	261 64.4	49 12.1	95 23.4
"In general, studying nursing is enjoyable"	n %	333 82.2	51 12.6	21 5.2
"It would be better if the present English language courses include exercises on English-English dictionaries and how to use them"	n %	323 79.7	53 13.0	29 7.1

KORREKTUR:

Seite IV: CIP:

Diab, Turki: Pedagogical Lexicography :
a case study of Arab nurses as dictionary users
/Turki Diab - Tübingen : Niemeyer, 1990
(Lexicographica : Series maior; 31)

Umschlag-Seite 4:

Turki **Diab**: Pedagogical Lexicography.
A Case Study of Arab Nurses as Dictionary Users.

FSC
www.fsc.org
MIX
Papier | Fördert gute Waldnutzung
FSC® C083411

Kolibri 360 GmbH
Ferdinand-Jühlke-Straße 7
99095 Erfurt, Deutschland
produktsicherheit@kolibri360.de